# ACOUSTIC WAVES IN BOREHOLES

Frederick L. Paillet
Chuen Hon Cheng

CRC Press
Taylor & Francis Group
Boca Raton London New York

CRC Press is an imprint of the
Taylor & Francis Group, an **informa** business

CRC Press
Taylor & Francis Group
6000 Broken Sound Parkway NW, Suite 300
Boca Raton, FL 33487-2742

First issued in paperback 2020

ISBN 13: 978-0-367-58000-1 (pbk)
ISBN 13: 978-0-8493-8890-3 (hbk)

**Visit the Taylor & Francis Web site at
http://www.taylorandfrancis.com**

**and the CRC Press Web site at
http://www.crcpress.com**

**Library of Congress Cataloging-in-Publication Data**

Paillet, F. L.
    Acoustic waves in boreholes / Frederick L. Paillet, Chuen Hon Cheng.
        p. cm.
    Includes bibliographical references and indexes.
    ISBN 0-8493-8890-2
    1. Oil well logging, Acoustic. I. Cheng, Chuen Hon Arthur. II. Title.
TN871.35.P35  1991
6221'.1828—dc20                                                                91-26673
                                                                                    CIP

Library of Congress Card Number 91-26673

# Preface

Borehole geophysics or well logging has been a highly specialized technology in the petroleum business for more than half a century. During that time, equipment and data processing methods have evolved, but the basic approach has remained fairly consistent: calculating hydrocarbon saturation from the appropriate combination of electric and porosity logs. Acoustic logging was developed in an effort to expand the arsenal of alternative porosity logs at the geophysicist's disposal. In that sense, acoustic porosity logging appeared to be an open-and-shut case. One simply measured the time required for acoustic waves to travel along a given length of formation adjacent to the borehole, and then inverted for velocity or transit-time. Interpretation focused on the practical problem of when and under what conditions the rock "matrix" velocity remained relatively constant so that one could correlate reductions in velocity with increases in fluid-filled porosity. Other applications for acoustic logs came almost as an afterthought. Variations in velocity could, of themselves, be significant, as in the detection of abnormal fluid pressures. More ambitious attempts to relate acoustic logs to the broad-scale velocity contrasts recognized in surface seismic reflections surveys were not very satisfactory. Identification of seismic reflections in well logs depended more upon intuition and experience than anything else. As a result, interpretation of porosity from transit time measurements remained the lone quantitative application of acoustic well logs up until the 1970s.

All of this changed after 1970. A number of factors combined to generate interest in what were known as "full waveform" acoustic logs — logs obtained by digitizing the complete acoustic pressure signal rather than simply detecting acoustic arrivals. These logs were displayed as seismic sections extending along the length of the borehole. Analysts used the same qualitative or semiquantitative methods employed in other seismic interpretation methods. But the application of specific, quantitative methods proved dishearteningly difficult. Even such straightforward tasks as the picking of shear arrivals or estimation of intrinsic attenuation proved more difficult than initially anticipated. Moreover, the methods yielded results in a seemingly unpredictable fashion; sometimes they worked reasonably well, and sometimes they did not work at all. The disillusionment in such potentially important areas of application as fracture characterization and mechanical properties determination was beginning to become apparent as the energy crisis reached its climax in the later 1970s.

We now know that the initial disappointments of full-waveform logging were the results of a familiar pattern in geophysics — the oversimplification of an inherently complicated problem. The supposedly simple body waves traveling along fluid-filled boreholes represent a much more complex wave propagation situation than do body waves in an infinite solid. The observed full-waveform logs represent the pressure variations in a fluid only indirectly related to the wave motions in the surrounding rock. The primary waves themselves represent complex couplings between fluid and wall rocks, sensitive to such factors as wall roughness and alteration of rock minerals in contact with borehole fluids. Yet out of such complexity arise new and important applications. The dependence of observed wave properties on so many different parameters implies the ability to interpret rock properties in greater detail than originally expected. The complexity of the interdependence also offers new opportunities to engineer geophysical equipment to measure properties of interest. As a result, the initial plateau of disappointment has been followed by new heights of interest and potential applications extending beyond those originally imagined.

The renewed interest in geophysics and the increasing technology of well logging has motivated many recent geophysical logging references. Some of these new books are exclusively concerned with petroleum related applications, while others address a more general geotechnical readership. Why is this an appropriate time for the publication of a monograph related to the narrow subject of acoustic waveform logging? We can provide two specific answers. First, the complicated theory for the basic wave propagation problem is now virtually complete. All important details, including the geometric radiation of wave energy, mode partition of source energy, effects of casing and cement bonding, and general character of borehole response for the full range of natural seismic properties appear to be in place. At the same time, the active areas of waveform log applications are rapidly expanding. The emergence of reliable, quantitative results from waveform log analysis has generated great interest in characterizing the mechanical properties of petroleum reservoirs with greater precision than ever before. Such results permit design of advanced recovery treatments that might otherwise not be possible. These same methods have much greater potential in the areas of mining engineering and construction, where they have just begun to be used. The advanced application of vertical seismic profiles and high resolution seismic provides other important applications for full waveform logs in bridging the gap between surface surveys and laboratory measurements of the properties of recovered samples. The intense interest in potential deep aquifer pollution, radioactive waste disposal, hydrocarbon storage facilities, and investigation of potential seismic hazards insure that the recently emerged wave propagation theory will find continued use, and perhaps entirely new areas of application, in the coming decade.

We expect full waveform log analysis to expand and evolve over the coming years. Interpretation methods are rapidly improving today, and we expect them to continue to improve in the future. This book is written to provide the fundamental information on the borehole wave propagation problem that will be required by scientists and engineers in the future. Primary emphasis is placed on physical mechanisms and on fundamental relationships between wave propagation and such physical properties as rock velocities, borehole conditions, and geometrical factors. The first two chapters outline the general history of seismic investigation, showing how borehole measurements have been integrated into large-scale investigations and reviewing conventional sonic porosity logging. The third chapter concentrates on the fundamental wave propagation problem, using a simplified, two-dimensional analog to the cylindrical problem. This approach provides insight into the ways that the physical properties of the fluid and surrounding rock influence the amplitude and frequency of waves refracted along the borehole wall. A subsequent chapter then treats the full mathematical problem. The next three chapters explore the dependence of waveform logs on rock properties and borehole conditions using specific examples. These include borehole wall alteration or invasion, casing, and conditions of cement bonding. The presentation is divided into open hole (Chapters 5 and 6) and cased hole (Chapter 7) subsections. The calculations and field data illustrate the effects of borehole size, source frequency, seismic velocities, Poisson's ratio, receiver spacing, intrinsic attenuation, and other factors on observed waveform characteristics. The final two chapters address the important problem of permeability estimation in fractured and unfractured rocks. We hope that this book will serve as the basic reference for waveform log investigations, while stimulating development of entirely new waveform log applications based on the insights we have provided.

*This book is dedicated to J. E. White in honor of his contributions to the theory and practice of borehole seismology. His personal guidance as teacher, colleague, and mentor has been an important part of our professional development in geophysics and has gone a long way towards making this book possible.*

# The Authors

Frederick L. Paillet, Ph.D. is Chief, Borehole Geophysics Research Project, U.S. Geological Survey, Lakewood, Colorado.

Dr. Paillet received his B.S., M.S., and Ph.D. degrees in 1968, 1969, and 1974, respectively, all from the Department of Mechanical and Aerospace Sciences, University of Rochester, New York. After completing military service with the U.S. Air Force Flight Dynamics Laboratory, Wright-Patterson Air Force Base, he was appointed Assistant Professor, Department of Geology, Wright State Universtiy, Dayton, Ohio. He Joined the U.S. Geological Survey in 1978 and became Research Project Chief in 1983.

Dr. Paillet is a member of the Society of Professional Well Log Analysts, the American Geophysical Union, the Society of Exploration Geophysicists, the Geological Society of America, the American Meteorological Society, the American Association for the Advancement of Science, and the Minerals and Geotechnical Logging Society. He is currently Associate Editor for Acoustic Logging, *The Log Analyst.*

The Society of Professional Well Log Analysts honored him with its Best Paper Award at its 1980 Symposium; he also served as the society's Distinguished Speaker from 1987 to 1988. He has published more than 50 research papers. His current research interests include acoustic logging, *in situ* interpretation of the geomechanical properties of rocks, ground water flow in fractured rocks, scale effects in geomechanical problems, and the application of geophysical inversion theory to well logs in hydrogeology.


Dr. C.H. (Arthur) Cheng is a Principal Research Scientist and the Director of the nCUBE/ERL Geophysical Center for Parallel Processing at the Earth Resources Laboratory in the Department of Earth, Atmospheric, and Planetary Sciences at the Massachusetts Institute of Technology in Cambridge, Massachusetts.

Dr. Cheng received a B.Sc. in Engineering Physics from Cornell University, Ithica, New York in 1973 and a Sc.D. in Geophysics from M.I.T. in 1978. He has remained with M.I.T., becoming a Principal Research Scientist in 1983. Since 1982 he has been the Project Leader of the Borehole Acoustics and Logging (formerly known as the Full Waveform Acoustic Logging) Consortium at ERL.

Dr. Cheng is a member of the Society of Exploration Geophysicist, the American Geophysical Union, the Society of Professional Well Log Analysts, and the Acoustical Society of America. He has organized or co-organized workshops in Parallel Computing in Geophysics, Seismic Tomography, Fluid

Flow in the Crust, and the Role of Geophysics in Horizontal Wells for the Society of Exploration Geophysicists. He is also a Member of the SEG Council.

Dr. Cheng has been the recipient of research grants from the National Science Foundation, Department of Energy, and various petroleum industry companies. He has published over 50 papers, mostly on borehole geophysics. His current research interests are seismic wave propagation, tomography, borehole geophysics, and rock physics.

# Acknowledgments

Many individuals have made important contributions towards the results presented in this book. Perhaps the most important contributions include the many small technical achievements on the part of colleagues and students which, taken together, result in the broad range of theory and application described in this book. Over nearly a decade of active research in borehole acoustics, numerous graduate students and postdoctoral associates at the Massachusetts Institute of Technology, Earth Resources Laboratory have produced the various borehole seismograms computations and interpretations illustrated in some of the figures. Most notable of these are Mark Willis, Ken Tubman, Dan Burns, Denis Schmitt, Xiao Ming Tang, Lisa Block, Karl Ellefson, and Jeff Meredith. Wayne Pennington, Dan Burns, and Dan Moos provided critical reviews of the first draft of the manuscript, and in doing so contributed to the effectiveness of the final result. U.S. Geological Survey equipment design engineer Alfred Hess, and technicians Richard Hodges and William Bruns worked wonders in obtaining field data and fighting equipment problems under adverse conditions at a number of research sites. Cliff Davison and Nash Soonawala, Atomic Energy of Canada Limited, Whiteshell Nuclear Research Establishment provided the core analyses and other data used in the geotechnical interpretation of waveform logs obtained at the Underground Research Laboratory in Lac Dubonnet, Manitoba; AECL technicians Don Solberg and John Gorie were especially helpful in solving numerous logistical problems in the field. Finally, we express our appreciation to Ann and Winnie for their understanding and patience during all of those late nights at the office required to make this book possible.

At the Earth Resources Laboratory at M.I.T., the work has been supported by the Full Waveform Acoustic Logging Consortium, whose past and present members are: Amoco, Aramco, Atlas Wireline, Britoil, Chevron, Conoco, Dome, Elf Aquitaine, EnTec, Exxon, Geophysical Services Inc., Getty, Gulf, Halliburton Logging Services/Welex, Intevep, Japex Geoscience Institute, Marathon, Mobil, NL McCullough, Oyo, Phillips, ResTech, Schlumberger-Doll, Seismograph Services Co., Sohio/BP, Statoil, Sun, Texaco, Total, and Unocal. It has also been supported by individual contracts with Chevron, Mobil, GSI, and Welex. In addition, portions of the work have been supporteed by grants from the Department of Energy and the National Science Foundation. Administration support from Sara Brydges has been indispensible. Nafi Toksoz, the Director of ERL, had the foresight to form the laboratory in 1982 and provided the unique environment to conduct research. The authors started their cooperation when Fred Paillet was the first Visiting Scientist at ERL for a year from 1982 to 1983.

# Contents

# 1

# Acoustic Measurements in Boreholes

Acoustic measurements in boreholes have become an important research subject in geotechnical applications ranging from conventional acoustic logging in petroleum exploration to sophisticated borehole-to-borehole tomography in tunnel construction and mine engineering. For many years these measurements were made as one element within wider-ranging studies based on a number of different geophysical measurements. For example, conventional acoustic transit-time logs are often treated as one of several useful formation porosity logs, or velocity distributions determined from acoustic logs are used to calibrate surface seismic data. Acoustic well logs are discussed in several different texts and monographs in just this way. Pickett (1960), Guyod and Shane (1969), Hearst and Nelson (1985), and Tittmann (1986) reviewed acoustic logging and some of the elementary theory relating to acoustic measurements in boreholes. White (1983) presented one of the most detailed discussions of the theory of acoustic wave propagation along fluid-filled boreholes. The theory was presented both in the context of borehole measurements as an extension of surface seismic methods (i.e., check shots and depth calibration of travel times) and as a method for characterizing lithologies *in situ*. However, none of these excellent references discussed waveform interpretation in detail, and none provided illustrative examples of waveform calculations for a wide variety of rock types and formation properties.

This monograph is directed towards the rather narrowly focused topic of acoustic waveform logging in boreholes and the interpretation of full waveform acoustic logs. Our intent is to provide a concise but thorough review of the theory and experimental evidence for the current applications of waveform logging. The first chapter describes the fundamentals of acoustic logging and elastic wave theory relating to the characterization of rocks surrounding a fluid-filled borehole. We then go on to review the development of acoustic measurements in boreholes, describing conventional acoustic logs and the various components used to obtain and record acoustic waveform logs. The third and fourth chapters review the theory for synthetic microseismograms, first outlining the general physical mechanisms involved, and then deriving the Fourier solution integrals in a rigorous way. Starting with the fifth chapter, we use the synthetic microseismogram calculation to explore the effects of seismic velocities and borehole geometry on waveform logs. Subsequent chapters present the application of theoretical insight derived from synthetic microseismograms to such practical problems as shear velocity logging, attenuation measurement, fracture characterization, and permeability estimation. Synthetic microseismogram calculations are expanded to include cased boreholes, and the interpretation of waveforms obtained under well bonded and poorly bonded conditions. Carefully selected field examples are used to illustrate waveform applications in each of these specific areas.

## 1.1   GENERAL COMMENTS ON ACOUSTIC LOGGING AND WAVEFORM LOG INTERPRETATION

The original applications of acoustic logging were based on the concept of measuring formation compressional velocity, $V_p$, *in situ*. The first acoustic transit-time logging systems were designed to measure $V_p$ as a bulk property of the formation, just as other logging systems measure bulk density or electrical resistivity (Vogel, 1952; Summers and Broding, 1952). In geophysical logging it is generally recognized that geophysical data needs to be corrected or adjusted to account for the presence of the borehole. In the case of acoustic logging, it was assumed that the conventional acoustic logging measurement (Figure 1.1) needed little correction because the acoustic signal traveled directly through the borehole wall from one receiver to the next. The travel times of critically refracted compressional or shear waves between a pair of receivers located along the borehole axis can be given as the sum of travel times within the borehole fluid ($\Delta t_f$) and in the surrounding rock ($\Delta t_r$) (Baker, 1984):

$$\Delta t = \Delta t_r + 2\Delta t_f$$
$$\Delta t_r = \Delta z / V_p$$
$$\Delta t_f = R \sqrt{\left( V_p^{\,2} - V_p^{\,2} \right)} \Big/ V_p V_f, \qquad\qquad 1.1$$

$R$   = borehole radius (no tool),
or
$R$   = annulus width (tool surface to wall),
$\Delta z$ = vertical distance between the acoustic source and receiver.

Borehole compensation in acoustic logging was interpreted as the process of removing the effect of acoustic "fluid" delay from the velocity measurement process. That is, the early acoustic logging equipment designers used ray theory to trace critically refracted paths through the borehole fluid from transmitter to the borehole wall, and then from the borehole wall to each receiver.

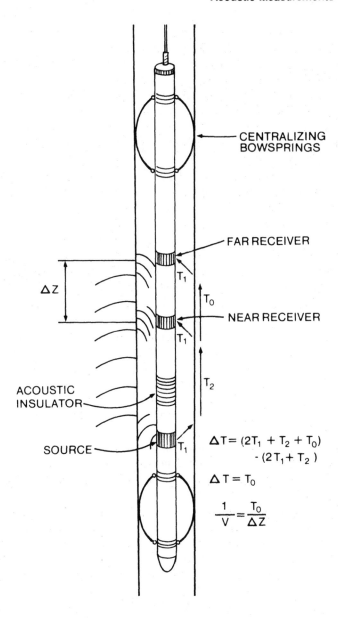

FIGURE 1.1. Schematic illustration of acoustic logging probe in a fluid-filled borehole, showing ray-tracing construction of acoustic travel time given in Equation 1.1.

If the borehole diameter does not vary between source and receivers, and the tool is centralized in the borehole as shown in Figure 1.1, then the difference between first arrival time at each receiver represents the time for acoustic energy to travel between receivers through the borehole wall. The subtraction of the near receiver acoustic travel time from the far receiver travel time removes the two fluid delays (from transmitter to wall, and from wall to receiver). This calculation does not fully remove the fluid delays if the borehole diameter changes between the two receivers. In order to treat such cases, acoustic logging equipment designers invented the "compensated" acoustic logging probe (Figure 1.2). This version of the probe has transmitters above and below the receiver pair so that errors attributed to changes in fluid delay between receivers are canceled by averaging the transit times measured for up- and down-going signals.

The use of compensated acoustic logging probes appeared to solve most of the problems in the measurement of *in situ* acoustic velocity. Such equipment made routine acoustic logging possible, so that commercial well logging companies could produce profiles of $V_p$ in boreholes as a standardized service. The profiles of compressional velocity seemed to be generated so easily and routinely that many researchers attempted to generate similar profiles for $V_s$ (shear velocity) from the same acoustic data. The approach was based upon detection of shear arrivals as abruptly increasing acoustic amplitudes at times in the measured pressure signal after the first compressional arrivals at the acoustic receivers (Willis, 1983; Aki et al., 1982). This approach assumed that the major obstacle to shear velocity profiling was development of sophisticated signal processing equipment capable of consistently recognizing first shear energy arrivals within the later part of the compressional arrival. Shear arrivals were readily identified in at least some waveforms when the pressure fluctuations at the receivers were monitored. It was assumed that advanced processing of these pressure signals would provide shear velocity logs in almost all formations.

Until approximately a decade ago, almost all acoustic logging research was focused on advanced application of data obtained from conventional $V_p$ profiles given by acoustic logs, or development of new data processing methods capable of producing similar profiles of $V_s$. In both cases, the approach was based on the implicit assumption that acoustic logging was a method for obtaining *in situ* measurements of body wave velocities as bulk properties of the formation. Interpretation methods would be based on the establishment of relations between the known properties of constituent minerals and fluids, and the measured seismic velocities of the composite rock formed by these constituents.

Although early papers by Biot (1952) and White and Zechman (1968) presented all of the formal theory required to calculate synthetic borehole microseismograms, those equations were not extensively used to investigate the properties of acoustic waves in boreholes until after the more recent studies by Rosenbaum (1974), Peterson (1974), Tsang and Rader (1979), and Cheng and Toksöz (1981). When these investigations into the properties of synthetic microseismograms had become well known, log analysts began to realize that acoustic logs were measuring something other than simple body waves. The

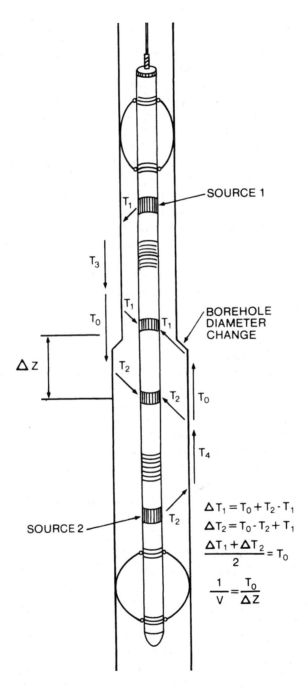

FIGURE 1.2. Schematic illustration of acoustic source and receiver configuration in compensated acoustic logging probe.

analysis demonstrated that the measured arrivals were the first in a series of critically refracted and trapped modes traveling along the borehole. The numerical results confirmed that although signals measured by conventional acoustic and advanced shear logging systems were traveling at the appropriate velocities, they were far different from "simple body waves" (Tsang and Rader, 1979; Paillet and Cheng, 1986). Although this analysis greatly complicated the interpretation of acoustic logs, the complexity of the theory added several advantages. First, the calculations showed that certain frequencies (depending upon borehole diameter and other conditions) were most effective in exciting the desired waves (Paillet and White, 1982; Paillet and Cheng, 1986; Schmitt and Bouchon, 1985). At the same time, calculated waveforms were shown to be dependent upon a number of formation parameters and borehole properties other than just seismic velocity. Most of these dependencies were shown to apply for certain frequency ranges and borehole conditions. At least some of these results have opened up entirely new applications for borehole acoustics.

The latest results in acoustic logging serve to reunite the two divergent approaches represented by seismic velocity profiling, and by use of full waveform logs in which the entire pressure signal is recorded to investigate borehole conditions such as extent of fracturing and quality of cement bonding. This monograph is based on the assumption that the two approaches will continue to converge and expand as the ability to model and interpret acoustic data in boreholes improves. The critical element in the sudden increase of applications for borehole acoustics appears to have been the sound understanding of the wave propagation problem made available through the synthetic microseismogram calculations and the verification of those calculations with carefully designed experiments. Even as the technology continues to advance at a rapid pace, the development of new applications for acoustic waveform logs will be based upon a thorough understanding of the wave propagation problem. Our primary intent is to document and demonstrate by example all of this fundamental understanding under the assumption that such insight will remain useful even as the technology of today becomes obsolete.

## 1.2    SEISMIC PROPERTIES OF ROCKS AND THEIR CHARACTERIZATION

Continuous, solid materials such as rocks can be represented as elastic bodies with a linear relationship between stresses and displacements. Numerous standard references on solid mechanics (Timoshenko and Goodier, 1951; Love, 1944) give the derivation of this linear relationship, showing that the properties of isotropic and homogeneous elastic solids can be specified by the two Lamé constants, $\lambda$ and $\mu$. The same references also demonstrate that the dynamic equilibrium conditions in the interior of such a solid result in the propagation of two types of elastic waves outward from the point at which an initial displacement is applied. These two kinds of waves are the compressional and shear body waves traveling at the two characteristic velocities, $V_p$ and $V_s$:

$$V_p^2 = (\lambda + 2\mu)/\rho$$
$$V_s^2 = \mu/\rho$$

$$1.2$$

where $\rho$ = density of the elastic solid.

Although the two Lamé constants are convenient for the mathematical description of solids, they are difficult to measure directly in the laboratory. Instead, we measure the elastic properties of rocks and other materials by compressing a rectangular or cylindrical sample in a press in the laboratory. In one simple test, a compressive stress is applied along one axis of a sample, and the sides of the sample are left stress-free, allowing lateral expansion. The stress applied to the sample is then given by the total force applied divided by the cross sectional area, and the strain is percent shortening of the sample under compression. The ratio of the applied stress to the measured strain defines the Young's modulus, $E$. A second characteristic property is obtained by measuring the percent increase in the lateral dimension of the sample during the same compression test. The ratio of lateral to longitudinal strain is the Poisson's ratio, $v$. There are no strict limitations on values for the Young's modulus, but the Poisson's ratio for an isotropic material is restricted to the range from 0.00 (no lateral expansion in compression) to 0.50 (liquids). Synthetic composite materials have been developed with negative Poisson's ratios. However, most natural minerals and sedimentary rocks composed of those minerals have values of n ranging from 0.20 (poorly cemented sandstones) to 0.45 (plastic shales) (Tathman, 1982; Wilkens et al., 1984; Castagna et al., 1985). Seismic velocities, $V_p$ and $V_s$, are related to the Young's modulus and Poisson's ratio by the equations (White 1983)

$$V_p^2 = (1 - v)E/\rho(1 - v - 2v^2)$$
$$V_s^2 = E/2\rho(1 + v)$$

$$1.3$$

Note that computation of seismic velocities from the elastic constants of rocks requires independent information on $\rho$, the bulk density. It also turns out that the elastic constants measured in the laboratory during compression tests can be significantly smaller than those inferred from seismic velocities (Cheng and Johnston, 1981; Bowles, 1978). The difference between "static" and "dynamic" elastic moduli appears to be accounted for by relaxation phenomena that allow a slow, transient decrease in stress or increase in strain after the initial application of the force (King, 1969; Spencer, 1981).

The conservation of energy requires that wave amplitudes decrease during outward propagation because of the geometric spreading of wave energy as the area of the entire wavefront expands. However, all rocks exhibit an additional amount of intrinsic attenuation attributed to as yet undetermined dissipational mechanisms that convert elastic wave energy to heat. It has been found by experiment that these losses often are nearly proportional to frequency, or that individual Fourier components of wave energy experience the same fractional attenuation traveling the same number of wavelengths. The linear dependence of attenuation on frequency is found to apply over the frequency range used in seismic exploration and acoustic logging (10 Hz to 50 kHz). Thorough reviews of theories accounting for the observed intrinsic attenuation in rocks and their mathematical expressions are given by Futterman (1962), Johnston (1978), Kjartansson (1979), Johnston and Toksöz (1981), and White (1983). The one prominent exception to the linear proportion between attenuation and frequency is the case of fluid-saturated porous rocks where the differential motion between mineral and fluid introduces additional dissipation (Biot, 1956a; 1956b). In that case, a maximum in attenuation occurs at a frequency that depends on the permeability of the sample. The experimental identification of local maxima in seismic attenuation with porous fluid-filled rocks, or rocks containing a mixture of fluid and gas, has become an important research topic in petroleum exploration (Dutta and Ode, 1979a, 1979b, 1983; Spencer, 1981; Murphy, 1984).

One of the most important effects of intrinsic attenuation is the introduction of velocity dispersion. A detailed investigation of the consequences of intrinsic attenuation indicates that seismic velocities increase with increasing frequency (Futterman, 1962; Strick, 1970; Kjartansson, 1979; Aki and Richards, 1980). If the spacewise attenuation of a wave propagating at the reference frequency $\omega_0$ is expressed as

$$A\Big|_{x=x_0} = A\Big|_{x=0} e^{-\alpha x 0} ; \alpha = \alpha(\omega)$$

then the velocity dispersion can be represented approximately (Azimi et al., 1968; Aki and Richards, 1980):

$$\frac{V(\omega)}{V(\omega_0)} = 1 + \frac{1}{\pi Q} \ln\left(\frac{\omega}{\omega_0}\right)$$

where $\quad Q = \dfrac{1}{2V_\infty}$, and $V_\infty = \lim_{\omega \to \infty} V(\omega)$

1.4

This increase in seismic velocities with frequency has been verified many times by experimental measurements over the range from 100 Hz to 100 kHz.

However, $V_p$ cannot continue to increase indefinitely with frequency beyond 100 kHz, but must approach a finite limit denoted by $V_\infty$ to satisfy the causality criteria given by Kjartansson (1979). A practical modification of Equation 1.4 gives the ratio of velocities $V_1$ and $V_2$ at two different frequencies, $\omega_1$ and $\omega_2$:

$$\frac{V(\omega_1)}{V(\omega_2)} = 1 + \frac{1}{\pi Q} \ln\left(\frac{\omega_1}{\omega_2}\right) \qquad 1.4A$$

The consistency with which intrinsic attenuation in naturally occurring rocks can be expressed as linearly proportional to frequency has resulted in the nearly universal use of the symbol "$Q$" to represent attenuation. The formal definition of $Q(\omega)$ is given as

$$\frac{1}{Q(\omega)} = -\frac{1}{2\pi}\frac{\Delta E}{E} \qquad 1.5$$

where $\Delta E$ is the loss of strain energy in one cycle of wave oscillation, and $E$ is the peak strain energy density stored during the cycle. In terms of wave amplitude, $Q$ is defined as

$$\frac{1}{Q(\omega)} = -\frac{1}{\pi}\frac{\Delta A}{A} \qquad 1.5A$$

where $\Delta A$ is the decrease in amplitude over one cycle, and $A$ is the maximum amplitude during the cycle. In general, different values of $Q$ are found for compressional and shear waves in the same elastic solid. Compressional and shear attenuation are denoted by $Q_p$ and $Q_s$; in situations where attenuation refers to wave motion representing a combination of wave modes, attenuation is denoted by $Q$ without subscript. Although $Q$ is formally defined as a function of frequency, $Q$ is usually assumed to be independent of frequency over the frequency range of interest. This assumption is justified by experiment in that measurement errors in $Q$ are almost always larger than the expected frequency variation over the accessible frequency range (Tittman et al., 1980, 1981; Winkler, 1986). For virtually all acoustic waveform applications, attenuation can be represented by the three constants $Q_p$ and $Q_s$ (formation compression and shear attenuation) and $Q_f$ (compressional attenuation in the borehole fluid).

The constant $Q$ approximation for intrinsic attenuation in rocks appears well-founded, but some questions have been raised over the extension of this characterization to the borehole fluid (Schoenberg et al., 1987). In much of the elastic wave propagation theory presented in this monograph, the borehole fluid will be treated as an elastic solid in which $V_s = 0$. However, viscous attenuation in water or drilling mud may introduce important effects that need to accounted for. Schoenberg et al. (1987) demonstrated that the importance of the viscous boundary layer for oscillations within the borehole is a function of the ratio of the borehole diameter to the thickness of the viscous boundary layer:

$$\delta = \sqrt{\frac{2\eta}{\omega\rho_f}} \qquad\qquad 1.6$$

where $\eta$ is viscosity and $\rho_f$ is the bulk density of the borehole fluid.

For water and most drilling fluids, and over the range of frequencies employed in acoustic logging, this ratio is always small, and the effects of fluid viscosity can safely be neglected. A similar conclusion was reached by Burns (1988) in a more thorough analysis.

## 1.3    LABORATORY MEASUREMENT OF SEISMIC VELOCITY AND ATTENUATION

Seismic velocities can be measured in the field using explosive sources and geophone arrays as in conventional surface refraction and reflection studies (White, 1983; Aki and Richards, 1980). However, the velocities inferred from such investigations are obtained using frequencies far below those used in geophysical logging, and they average rock properties over large volumes of rock. The seismic velocities of smaller rock volumes, closer in size to those investigated by acoustic logging, can be obtained by using laboratory equipment to measure the velocities of core samples (Figure 1.3). In this case, the volume of investigation is somewhat smaller and the applied frequencies somewhat higher than frequencies used in logging (100 kHz or more in laboratory tests vs. 10 to 20 kHz in logging; Toksöz et al., 1976; Toksöz and Johnston, 1979). The lower frequencies and larger volume of investigation in logging averages velocities over larger volumes, while the higher laboratory frequencies result in greater measured velocities than those given by logs according to the velocity dispersion in Equation 1.4. The sample volume difference can be surmounted by performing a large number of tests on core samples from adjacent intervals, and then spatially smoothing the resultant velocity distribution. A similar process is almost always recommended when comparing geophysical logs to core data as part of the common practice of using core data to calibrate logs. Depth matching is another important source of error. There will generally be a finite offset between the nominal core depths and the depths on the geophysical logs. This difference is attributed to differences in zero point reference, cable stretch, and measurement error. Core data needs to be smoothed and depth-correlated with the vertical adjustment required for maximum correlation before logs are calibrated.

In addition to possible depth errors and the difference in the scale of investigation, comparison of laboratory measurements of acoustic velocity of

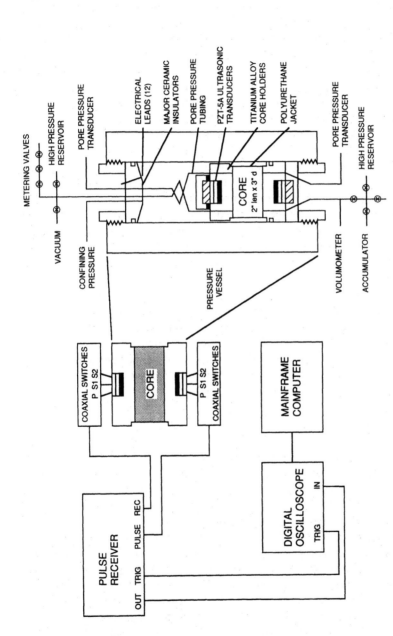

FIGURE 1.3.   Diagram showing equipment for measurement of the seismic velocities of rock samples in the laboratory under controlled confining and pore pressures.

core samples and acoustic log data can be complicated by a number of other factors. One of these is the potential alteration of core samples during and after recovery. Some lithologies are subject to disintegration during coring, so that samples from these rock types cannot be tested in the laboratory. Other samples may be recovered intact, but are subject to desiccation during storage, or hydration when exposed to water in the drilling fluid. Extreme care must be taken in handling cores from rocks composed of sensitive clay minerals which are subject to irreversible changes on exposure to air or moisture. Even when core samples are recovered intact and remain unaltered, samples to be tested in the laboratory need to be trimmed and fitted with impermeable jackets. This preparation process sometimes results in damage to samples. Sample breakage during trimming, for example, may bias laboratory measurements of $V_p$ towards the strongest or least brittle samples. All of these factors indicate the potential for biased results to occur through the selective process of core sample recovery, transportation, storage, and preparation.

Even when velocity dispersion, depth errors, sample damage, and spatial averaging over larger sample volumes are taken into account, seismic velocities measured from core samples do not usually correspond to the velocities given by logs. The difference has been found to result from the effects of *in situ* environment (pressure and temperature) on the seismic velocity of rocks. Velocity studies of pressurized core samples indicate that the pressure produced by the weight of overlying rocks has a large effect on seismic velocities at depths less than 1000 m, but is not so important at greater depths (Handin et al., 1963; Brace, 1965). In the case of porous sedimentary rocks, the changes in velocity with pressure are found to depend upon the effective stress $\sigma_e$ (Hubbert and Rubey, 1959; Hughes and Cooke, 1953; Nur and Byerlee, 1971):

$$\sigma_e = \left( \sigma_0 - P_p \right) \qquad \qquad 1.7$$

where $\sigma_0$ is the average principal compressive stress, and $P_p$ is the pore pressure. Departures of velocity from the effective stress dependence are found to be relatively small, except in situations where stresses are highly non-isotropic (Banthia et al., 1965; King, 1966, 1968; Brace and Martin, 1968; Brace, 1972; Garg and Nur, 1973; Carroll, 1979).

Modern laboratory apparatus for the measurement of the elastic velocities of rock samples include pressure chambers, flexible jackets for samples, and equipment for separate control of fining and pore pressure. Examples of the variation in compressional ($V_p$) and shear ($V_s$) velocities with effective confining stress for two representative sedimentary rock types are given in Figure 1.4. The curves given in the figure indicate the substantial differences in velocity that can occur at depths less than a few thousand meters (Birch, 1960, 1961, 1966; Knutson and Bohor, 1963).

The variation in seismic velocity with confining pressure has been a subject of extensive investigation. Early studies indicated that rock deformation during the application of stress increased the contact area among individual grains in granular sediments (Fatt, 1958, 1959; Timur et al., 1971). However, experiments showed that the same increase in seismic velocities with increasing confining pressure occurred for crystalline rocks in which grain contacts do not deform (Brace et al., 1968; Nur and Simmons, 1969). Subsequent analysis showed that these increases in velocity could be accounted for by the closure of small cracks or "microcracks" distributed along grain boundaries within essentially all rocks (Berryman, 1980a, 1980b). Microcracks can be described as a continuous distribution of cracks of different sizes and different degrees of elongation (aspect ratio). Fracture shapes can vary from spherical pores (requiring large confining pressures to close) to thin, lenticular openings (easily closed under low stress). Several recent papers examined the seismic properties of rocks characterized by spectra of various pore sizes and shapes, and considered the ways in which the observed variation of seismic velocity with pressure can be used to infer the spectrum of microcracks (Simmons et al., 1974; Toksöz et al., 1976, 1979; Cheng and Toksöz, 1978, 1979; Mavko and Nur, 1978; Korringa et al., 1979; Berryman, 1980a,b; Walsh, 1981; Moos and Zoback, 1983).

Variation of seismic velocities with temperature is not nearly so important as with effective stress, but could become important at greater depths where pressure effects on velocity are less significant, or in geothermal regimes where geothermal heating from intrusions and geothermal fluid circulation may introduce significant temperature gradients. Extensive experimental data on the variation of seismic velocities with temperature at typical confining pressures for various sedimentary, igneous, and metamorphic rocks were given by Hughes and Cross (1951), Timur (1977), and Christensen (1982) (Figure 1.5). These results show that increasing temperatures decrease seismic velocities at most by a few percent over the temperature range from 0 to 100°C, with the exception of some shales where temperature cycles may cause irreversible changes to the mineral framework (Johnston, 1987). Such changes may be caused by the effects of temperature on the moduli of rocks, but would be at least partially compensated by the decrease in density associated with thermal expansion. Timur (1977) suggested that thermal activation of microfractures may account for the observed decrease in velocity when superimposed on the changes in rock matrix. In any event, the relatively small changes in velocity observed for large increases in temperature for most rocks indicate that thermal effects on seismic propagation can be ignored in most practical applications.

Attenuation of seismic waves in rock samples also can be measured in the laboratory, often using the same equipment as in velocity measurements. Detailed discussions of the equipment and experimental techniques used in the laboratory measurement of $Q$ were given by Wyllie et al., (1962), Gardner et al., (1964), Toksöz and Johnston (1981), Winkler and Plona (1982), and White (1983). Most attenuation measurements are made by one of three different methods: (1) the measured amplitude decay of pulses transmitted through samples, (2) the characteristics of frequency resonance in bars, and (3) the

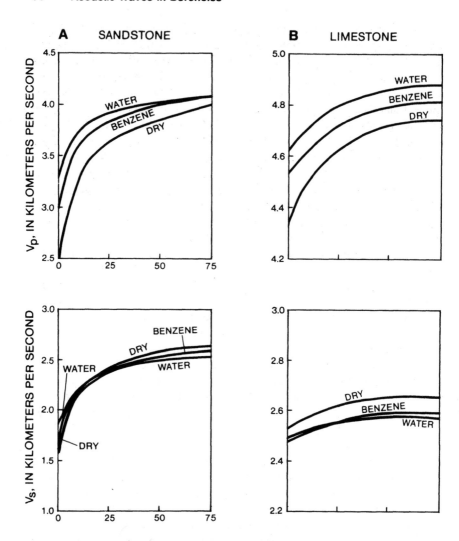

FIGURE 1.4. Examples of laboratory measurements of compressional and shear velocity under dry, water saturated, and benzene saturated conditions for (A) sandstone and (B) limestone.

amplitude decay of multiple reflections in bars. White (1983) and Toksöz and Johnston (1981) gave an excellent comparative review of the different measurement techniques. The first method is readily adapted for use in pressure vessels where the state of stress and environment of the sample can be controlled. However, care must be taken to account for the effects of beam spreading and reflections from sample edges in these measurements (Tang et al., 1990). Toksöz et al. (1979) gave measurements of attenuation in typical sedimentary rocks as a function of effective stress and saturation. Representative examples of the effect of increasing pressure on attenuation (inverse of $Q_p$ and $Q_s$) are given in Figure 1.6. Both $Q_p$ and $Q_s$ are found to increase (decreas-

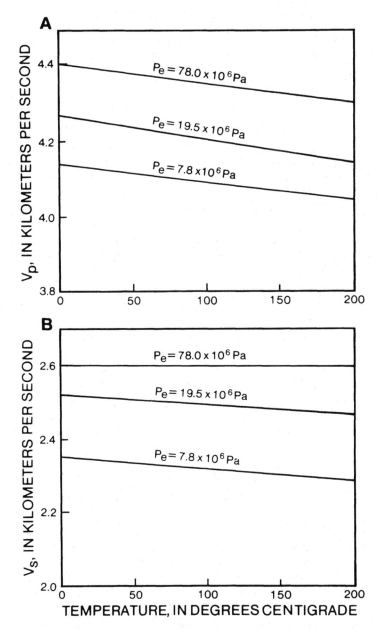

FIGURE 1.5. Variation of seismic velocities of a sandstone core sample in the laboratory under various effective stresses over the temperature range from 0 to 200°C (data from Timur, 1977).

ing attenuation) with confining pressure in a manner similar to the $V_p$ and $V_s$ (Johnston, 1978; Johnston et al., 1979; Johnston and Toksöz, 1980). Numerous experiments indicate that $Q$ is relatively large for kiln-dried samples, but decreases significantly after partial saturation with water or other fluids (Mann and Fatt, 1960; Kuster and Toksöz, 1974a,b; Nur and Byerlee, 1974, 1977; Mavko, 1979; Pandit and King, 1979; Spencer, 1979; Clark et al., 1980; Tosaya and Nur, 1982; Khilar and Fogler, 1983; Jones, 1986). This observation has important implications for theories about attenuation mechanisms, but is beyond the scope of this monograph (Winkler and Nur, 1979a,b, 1982).

## 1.4  GEOMETRIC EFFECTS AND SCALE OF INVESTIGATION

The geophysicist is constantly aware of the differences between homogeneous and isotropic elastic media considered in the continuum theory of classical mechanics and the natural mineral composites that form real rocks. The foremost consideration is the extent to which geological formations correspond to elastic continua. Geological formations often appear to be continuous media on scales much smaller than the acoustic wavelength of interest. Mineral or fluid inclusions and other inhomogeneities much smaller than the seismic wavelength are unimportant except to the extent that they contribute to the average properties of the bulk medium. Inclusions equal to or larger than the seismic wavelength scatter and refract seismic waves in a measurable way that can be directly related to the size and shape of the inclusion. One of the most important consequences of this scale effect is the complication introduced in the comparison of acoustic measurements made at very different scales of investigation.

The small-scale properties of geologic formations most often affect acoustic measurements in boreholes and their applications in three ways: (1) bulk porosity and permeability introduced by the distribution of fluid inclusions, (2) bulk anisotropy introduced by layering or lamination at subwavelength scales, and (3) differences in bulk velocities or velocity contrasts caused by slightly larger scale layering which degrade measured signals because of multiple reflections, etc. Porosity and permeability are two of the most important properties of rocks for almost all applications. The theoretical and empirical relationships between bulk porosity or permeability and measured seismic velocities have been a subject of intense study for many years, and are the subject of much of the remainder of this book.

The nonisotropic nature of many sedimentary rocks is evident from the finely laminated structure of core samples, and from differences in arrival times associated with different travel paths measured in various laboratory, surface seismic, or vertical seismic profile studies (White and Angona, 1955; White and Sengbush, 1953; White, 1983; White et al., 1983; Lo et al., 1986). Laminated sediments are modeled as transversely isotropic solids, in which seismic velocities have a vertical axis of symmetry. Postma (1955), White and Tongtaow (1981), White (1982, 1983), and Tongtaow (1982) described the mathematical formulation of the stress-strain relationship for transversely isotropic solids. In that case, seismic waves propagating along the borehole travel along the axis of symmetry. White (1983), Chan and Tsang (1983), and Crampin (1984) extended the analysis of Love (1944) to the case of a laminated solid to give seismic velocities for the composite solid for waves traveling across bedding planes:

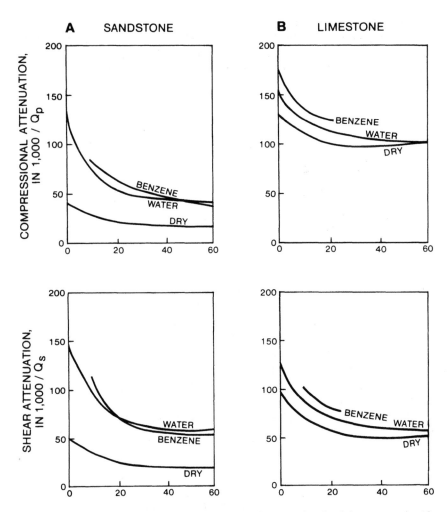

FIGURE 1.6. Examples of laboratory measurements of compressional and shear attenuation ($Q_p$ and $Q_s$) under dry, water saturated, and benzene saturated conditions for (A) sandstone, and (B) limestone.

$$V_p^{\,2} = \left( \sum_{i=1}^{N} \frac{C_i}{\lambda_i + 2\mu_i} \right)$$

$$\rho V_s^{\,2} = \left( \sum_{i=1}^{N} \frac{C_i}{\mu_i} \right)^{-1}$$

1.8

where $\rho = \sum_{i=1}^{N} C_i \rho_i$, and $\sum_{i=1}^{N} C_i = 1$

and $C_i$ is the volume fraction of layer $i$ with Lamé constants $\lambda_i$ and $\mu_i$.

In the formal mathematical treatment of transversely isotropic solids, the elastic properties of the solid are expressed by five elastic constants rather than the two Lamé constants required for the isotropic elastic case. If the Z axis is taken as the axis of symmetry, then the stress-strain relations take the form (Love, 1944)

$$\sigma_{xx} = A\varepsilon_{xx} + (A - 2N)\varepsilon_{yy} + F\varepsilon_{zz}$$
$$\sigma_{yy} = (A - 2N)\varepsilon_{xx} + A\varepsilon_{yy} + F\varepsilon_{zz}$$
$$\sigma_{zz} = F\varepsilon_{xx} + F\varepsilon_{yy} + C\varepsilon_{zz}$$
$$\sigma_{xy} = N\varepsilon_{xy} \qquad\qquad\qquad 1.9$$
$$\sigma_{yz} = L\varepsilon_{yz}$$
$$\sigma_{zx} = L\varepsilon_{zx}$$

where $\sigma_{ij}$ and $\varepsilon_{ij}$ are the stresses and strains in a transversely isotropic solid, and $A, C, F, L,$ and $N$ are the five elastic constants required to express the stress-strain relationship. These more complicated relations cannot be resolved into independent compressional and shear wave modes when substituted into the equations of motion for an elastic solid. Instead, we have modes designated "quasi-P" and "quasi-S" in which the anisotropy introduces a small coupling of shear vibration to the compressional wave, and a small coupling of compressional vibration to the shear wave. The shear coupling to the quasi-P wave, $a$, and the compressional coupling to the quasi-S wave, $b$, are given by (White, 1983)

$$a = \frac{M\left\{(F + 2L)\kappa^2 - AM^2 - \rho\omega^2\right\}}{i\kappa\left\{L\kappa^2 - (A - F - L)M^2 - \rho\omega^2\right\}}$$

$$b = \frac{i\kappa\left\{L\kappa^2 - (A - F - L)K^2 - \rho\omega^2\right\}}{K\left\{(F + 2L)\kappa^2 - AK^2 - \rho\omega^2\right\}} \qquad\qquad 1.10$$

where $\omega$ is the frequency, $\kappa$ is the z component of the wavenumber, and velocities of the quasi-P and quasi-S waves are given by

$$M^2 = \frac{-B + \sqrt{B^2 - 4AD}}{2A}$$

$$K^2 = \frac{-B - \sqrt{B^2 - 4AD}}{2A}$$

with

$$B = \left(p\omega^2 - Lk^2\right) + \left(p\omega^2 - Ck^2\right)A + (F + L)\kappa^2$$
$$D = \left(p\omega^2 - Lk^2\right)\left(p\omega^2 - Ck^2\right)$$
$$V_{qp} = \omega/M \qquad\qquad\qquad 1.10A$$
$$V_{qs} = \omega/K$$

There is also another shear wave involving transverse motion confined to the horizontal plane, the SH shear wave. If this wave travels at an angle $\gamma$ with the vertical, then the SH velocity is given by

$$V^2{}_{sh} = \frac{N\sin^2\gamma + L\cos^2\gamma}{\rho}$$

In general, the coupling of the compressional and shear motion applies to waves propagating in transversely isotropic solids. However, in the limit of wave propagation directed along or transverse to the axis of symmetry, the compressional and shear motions are found to uncouple. In the case of compressional propagation along bedding (perpendicular to the Z axis), we have $V_p = (A/\rho)^{1/2}$. In the case of shear waves traveling along bedding planes, we have $V_{sh} = (N/\rho)^{1/2}$ for "horizontal" motion (parallel to bedding and transverse to the axis of symmetry) or $V_{sv} = (L/\rho)^{1/2}$ for "vertical" motion across bedding planes. In borehole logging wave propagation is normally along the axis of symmetry, and $V_p = (C/\rho)^{1/2}$. Axisymmetric propagation along the borehole involves vibration that is mostly transverse to the Z axis or aligned with the bedding planes, so that $V_s = V_{sv} = (L/\rho)^{1/2}$. This is the value of shear velocity that we will most often be concerned with in acoustic logging. The one prominent exception will be in the case of the tube wave, the trapped borehole wave mode analogous to the Stoneley interface mode on the plane interface between a fluid and solid. When we deal with tube-wave propagation, the shear motion of the borehole wall involves radial displacements that distort the circumference of the borehole. In the low-frequency limit of tube-wave propagation, White (1983) showed that tube wave velocity depends upon the elastic constant $N$, and is therefore related to horizontally rather than vertically propagating shear waves.

When seismic velocity distributions determined from surface seismic surveys or vertical seismic profiles are compared to velocity profiles from well logs, the effects of scale of investigation, velocity dispersion, and anisotropy need to be considered (Lindseth, 1979; Strick, 1971; Thomas, 1978; Goetz et al., 1979). Velocity dispersion can be estimated from estimates of intrinsic attenuation. If the larger-scale seismic measurements are made from such large offsets that the direction of propagation is very different from the borehole axis, variation in seismic velocities with propagation direction also must be considered. However, Stewart (1984) found that the difference in scale introduces another effect related to the layering of natural formations. The greater vertical resolution of acoustic logs indicates velocity variation along the borehole when beds are more than 0.5 m in thickness. These beds cannot be resolved by the much lower frequency seismic waves from surface sources, but the impedance introduced by the layering can produce a significant delay in downward-propagating waves. Stewart et al. (1984) showed that such a delay is frequently observed in addition to the delay expected from velocity dispersion.

The most difficult problem in relating seismic measurements of rock properties conducted at very different scales of investigation is the characterization of fractured rock. Fractures act as discrete discontinuities for high-frequency seismic waves, but are averaged into bulk rock properties at larger scales (Stierman and Kovach, 1979). In the early literature, fractures were modeled as simple planar "sheets" or "slabs" of fluid embedded in elastic half-spaces (Snow, 1965; Hsieh et al., 1983; Davison, 1984). However, real fracture faces remain in contact through asperities that are deformed during compression (Witherspoon et al., 1981; Paillet et al., 1989). Contact of asperities gives fractures a mechanical strength which is intermediate between that of fluid and rock, and which is almost impossible to characterize using conventional methods. Rock properties also can be altered by geochemical processes during the passage of fluids, making the effects of alteration difficult to distinguish from the mechanical effects of fracturing (Keys, 1979; Moos and Zoback, 1983; Goldberg et al., 1985; Paillet, 1985; Goldberg and Gant, 1988). The seismic characterization of fractured rocks is one of the most difficult problems in geomechanics and will be treated in detail in subsequent chapters.

# 2

# Acoustic Well Logs and Porosity Logging

Acoustic well logging initially found widespread use as a method for determining the distribution of porosity in sedimentary rocks. Equipment for acoustic logging was developed over a period of decades under the assumption that the logs being obtained were accurate representations of the compressional velocity profile along the borehole. Largely empirical development of equipment and data analysis methods produced logs that could consistently be related to porosity distributions and appeared to agree with other conventional porosity logs. In addition, there were no inherent restrictions on logging speed as with nuclear porosity logs, and none of the problems associated with handling of radioactive sources. Therefore, acoustic transit time or "sonic" logs became a common and relatively inexpensive porosity log for petroleum exploration and reservoir characterization. The primary limits on the use of acoustic logs were the inability to log in cased holes and the difficulty in calibrating logs at depths above 1000 m, where formation compressional velocities vary significantly with confining pressure in addition to porosity and lithology.

## 2.1  ACOUSTIC POROSITY LOGGING EQUIPMENT

Acoustic porosity logging equipment developed for deep well logging applications is illustrated in Figure 2.1. The logging system consists of a probe with one or more transmitters and several receivers (Guyod and Shane, 1969; Hearst and Nelson, 1985). The logging tool illustrated in Figure 2.1 contains only a single pair of receivers and would not be classified as a modern compensated tool. However, the sonde could be converted to a compensated tool by adding a second transmitter to the sonde at the top of the tool body, with uphole electronics and switching modified to make transit time measurements from both the top and bottom receiver (Figure 1.2). The diagram in Figure 2.1 illustrates switching required to pulse the transmitter many times per second, producing transmitted signals at the receiver pairs at regular intervals. The electrical circuitry illustrated in Figure 2.1 is required to amplify the two receiver signals, transmit the signals up the logging cable, compare the ampli-

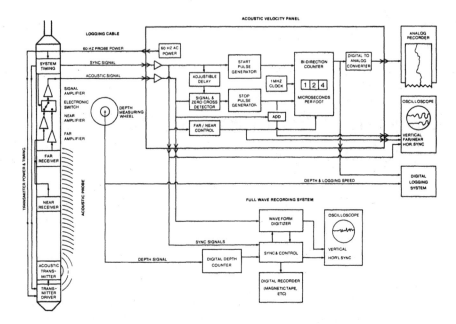

FIGURE 2.1.   Diagram showing electronics and related equipment for the measurement of acoustic transit times using the U.S. Geological Survey acoustic logging system.

fied signals to a preset threshold, and compute a measured transit time based on the time difference between when the threshold is exceeded at the far and near receivers. Operation of equipment such as that illustrated in Figure 2.1 requires considerable system tuning. For example, the acoustic signals at each receiver must be amplified by different amounts in order to compensate for the increasing signal attenuation with increasing distance from the source. The acoustic logging system uses built-in gains which provide amplitude ratios giving roughly equal amplitudes for various receiver spacings, with uphole gain controls so that the observed signals at the surface can be fine tuned for approximately equal amplitudes at all receivers.

When transit times are computed at the surface, acoustic signals relayed from the two downhole receivers are amplified so that first arrivals are very large. This procedure makes comparison with preset thresholds easy. The only limitation on this process is the potential amplification of noise, such as electronic noise in the system, and mechanical noise produced by the scraping of the logging sonde on irregularities in the borehole wall. These are not great problems because electronic noise is almost always negligible, and the logging tool has been designed with rubberized "bumpers" and padded centralizers so as to minimize "road noise". Various precautions are taken to dampen and delay acoustic signals that might be relayed along the high-velocity metal of the tool body. These include overlapping slots and flexible rubber hose sections between source and receiver. A more subtle problem is the effect of gain

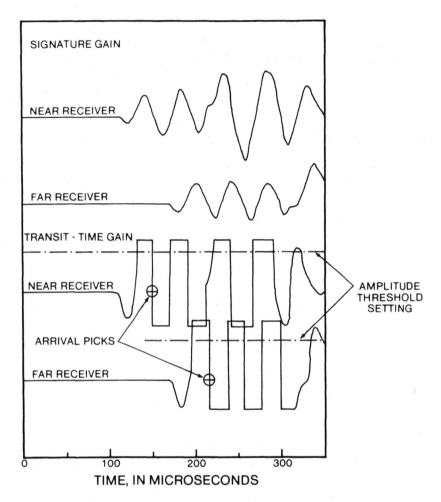

FIGURE 2.2. Illustration of signal amplification, amplitude threshold setting, and arrival time picking in acoustic data processing.

adjustments on the arrival time pick. Because uphole gains are adjusted arbitrarily to make the signal appear "good" at the surface panel, the exact arrival "pick" is made at an arbitrary point within the first quarter cycle of the first arrival (Figure 2.2). For this reason the arrival pick is postponed to take the first zero crossing or wave node after the amplitude threshold has been achieved, rather than picking at the amplitude threshold. Such a scheme produces a consistent phase at arrival-time picks for each receiver.

The transit time values computed at the wellhead in acoustic logging are determined by subtracting the arrival time at the near receiver from the arrival time at the far receiver, and then dividing by the receiver separation. This calculation assumes that the "fluid delay" for near and far receivers is the same

(Figure 1.1). These delays will be the same only if the logging tool is centered in the borehole, and only if the borehole diameter is constant over the length of the logging tool. A compensated acoustic logging tool is designed to at least partially account for changes in borehole diameter. The effects of borehole compensation are illustrated in Figure 1.2. An abrupt change in borehole diameter between the two acoustic receivers produces a difference in fluid delays for the near and far receivers. If a second acoustic source is symmetrically located above the two receivers as shown, the error associated with the diameter change will be exactly the opposite of that obtained by using the lower source. The error may therefore be removed by averaging the transit times determined using the upper and lower sources. Compensated acoustic logging tools use alternate firings of the upper and lower sources to produce interval transit times in which the first order effects of diameter variations are removed.

## 2.2   ACOUSTIC TRANSIT TIMES AND POROSITY LOGGING

Acoustic transit time or "sonic" logging quickly found its way into the standard inventory of geophysical logs used in the petroleum industry. The acoustic logging system was identified as a reliable system producing depth profiles of the inverse of acoustic velocity in the same way that other logging systems produce repeatable profiles of natural gamma activity or bulk density (Summers and Broding, 1952; Vogel, 1952). In many situations, the acoustic velocity of minerals such as quartz or limestone forming the matrix of porous reservoirs could be considered a constant at *in situ* temperatures and pressures. Then variations in interval transit time could be related to differences in porosity of the formation. According to these assumptions the measured transit time would be somewhere between the relatively short transit time of the mineral and the much longer transit time of the fluid filling the pore spaces. Strictly empirical studies soon demonstrated that the effect of porosity on acoustic transit time in otherwise acoustically uniform rocks was given by a volume-weighted average of the constituent transit times, known as the Wyllie time-average equation (Wyllie et al., 1956, 1958):

$$\Delta T = (1 - \phi)\Delta T_r + \phi \Delta T_f$$

or

$$\Delta T = \Delta T_r + \phi\left(\Delta T_f - \Delta T_r\right)$$

2.1

where $\Delta T_r = 1/V_p$ and $\Delta T_f = 1/V_f$. Note that this relationship gives very different results from taking the volume-weighted average of the acoustic velocities and then inverting for transit times.

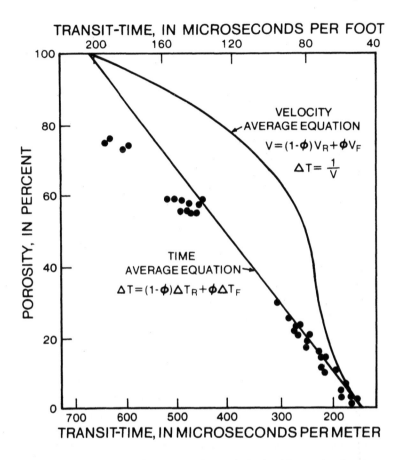

## TRANSIT-TIME, IN MICROSECONDS PER FOOT

VELOCITY
AVERAGE EQUATION
$$V = (1-\phi)V_R + \phi V_F$$
$$\Delta T = \frac{1}{V}$$

TIME
AVERAGE EQUATION →
$$\Delta T = (1-\phi)\Delta T_R + \phi \Delta T_F$$

POROSITY, IN PERCENT

TRANSIT-TIME, IN MICROSECONDS PER METER

FIGURE 2.3. Comparison of measured acoustic transit times with porosity data from core illustrating departure from the Wyllie time-average equation at porosities greater than 40% (data from Pickett, 1968).

Although the Wyllie time-average equation relating measured transit times to rock porosity has been in use for over three decades, this relationship is considered an empirical result. Under some restrictions, and using certain assumptions concerning the elastic properties of the mineral framework, one can derive a relationship between $V_p$ and porosity that has the same form as the Wyllie time-average equation (Geertsma, 1957; Geertsma and Smit, 1961). An example of the experimental data that support the Wyllie time-average equation is given in Figure 2.3. The figure illustrates that the volume average of transit times approximates real data more effectively than the volume weighted average of acoustic velocities. Also note that the data depart from the Wyllie time-average equation for porosities greater than 40%. It is generally found that the equation underpredicts transit times at larger porosities. One might think of the departure occurring because for large porosities the fluid-filled pores contribute to the slowing directly through the volume of acoustically slow fluid in the pore spaces, and indirectly through the weakening of the rock matrix.

Acoustic porosity logging provides an important tool for petroleum exploration and reservoir characterization where accurate porosity logs are in demand. The acoustic porosity log is relatively inexpensive to run, does not require the elaborate safety precautions associated with nuclear porosity logs, and there is no inherent limit on logging speed such as that imposed by nuclear counting rates and statistical considerations in neutron logging. However, there is one important complication in acoustic porosity logging that sometimes limits the accuracy with which porosities can be determined using acoustic transit time logs. That is the dependence of the acoustic velocity of rocks on *in situ* conditions. In the previous chapter we indicated the changes in velocity that accompany increased confining pressure. These natural velocity differences make it impossible to calibrate acoustic porosity logs where the measured transit times depend upon conditions unrelated to porosity, such as the large variation in $V_p$ with confining pressure.

Although the substantial changes in acoustic velocity of minerals with pressure appear to a major restriction on the use of acoustic porosity logs, practical experience indicates that these changes often become unimportant beyond a certain depth. The exact depth at which the effects of stress on velocity calibration can be ignored depends upon pore pressure and the deformability of the rock framework. Most, but not necessarily all, of the net stress exerted on the framework of a porous and permeable sedimentary rock is given by the difference between the compressive stress associated with the overburden and the fluid pressure exerted from within. In the case of normal geopressured reservoirs, the fluid pressure is somewhat less than the pressure that would occur at the base of a column of water extending to the surface. The total lithostatic stress would be equivalent to the weight of a similar unit column of saturated rock. Thus the total confining pressure ($P_c$) would be

$$P_c \approx \sigma_e = \left(\rho_s - \rho_f\right)\gamma d \qquad\qquad 2.2$$

where $\rho_s$ is the bulk density of the rock in the geologic column, $\rho_f$ is the density of the pore fluid, $d$ is depth, and $\gamma$ is a constant slightly less than 1.0 (usually about 0.9). For a typical sandstone saturated with water and the velocity variations given in Figure 1.4 the changes in acoustic velocity with depth become either negligible or small enough to remove with a linear correction at depths of burial exceeding 1000 m. However, this result does not apply in the case of abnormal geopressures encountered in sedimentary basins where rapid depth of burial and the low permeability of clay-rich sediments have not allowed for the escape of connate waters. In these situations it has been found that the observed departure of measured transit times from the expected decreasing trend with depth (increasing acoustic velocities) is a useful indicator of abnormal geopressures (Guyod and Shane, 1969).

The variation of acoustic velocities with confining pressure is important even in the case of hard crystalline rocks, as indicated by the velocity changes with confining pressure for granite and similar rocks by Simmons (1964), Nur and Simmons (1969), and Toksöz et al. (1976). Compressional velocities approach a constant value in large igneous intrusions at relatively shallow depths because elongated microcracks on grain boundaries close at low overburden pressures (Walsh, 1965; Spencer and Nur, 1976; Hadley, 1976; Feves and Simmons, 1976). This allows the fitting of observed transit time variations to a linear trend at relatively shallow depths in these formations. The observed variation in velocity with depth for an unusually homogenous granitic batholith located on the southwestern margin of the Canadian Shield is illustrated in Figure 2.4 (Paillet, 1980). The acoustic signals recorded at the two receivers show the gradual decrease in transit time for both shear and compressional arrivals with depth of measurement. Note that the very regular decrease in transit time could easily be represented by a linear equation, and the increase subtracted from the transit time measurements to remove the background trend.

## 2.3    RECEIVER GEOMETRY AND SOURCE CHARACTERISTICS OF POROSITY LOGGING EQUIPMENT

The general theory of acoustic measurements in boreholes does not place many restrictions on the exact choice of transducers and receiver spacing. However, experience has demonstrated that certain choices of source frequency and receiver offset are most effective. Acoustic logging sources in use are driven by either magnetostrictive or piezoelectric sources. In both cases, early logging tool design used sources that were constructed in such a way that an electrical forcing of the source transducer produced a mechanical vibration related to the natural resonant frequency of the source material. This approach ties source frequency to source size, with resonant frequencies decreasing as source dimensions increase. The source ringing produces a relatively narrowband signal, which may be modeled as a single harmonic modulated by a decaying exponential (Tsang and Rader, 1979; Cheng and Toksöz, 1981). Such sources work well in a scheme where arrival times are picked by identifying the first wave node after a preset amplitude threshold has been achieved, but would not be suited for visual picks of first energy arrivals.

Acoustic logging source transducers are designed as rings or cylinders aligned with the borehole axis so as to produce axisymmetric forcing. The signals produced in this manner may then be averaged around the circumference of cylindrical detectors some distance away along the body of the logging tool. This source and receiver configuration requires that the acoustic logging tool be centralized within the borehole. Centralization is accomplished using bowspring or rubber arm centralizers (Figure 1.1). However, borehole alignment and other factors almost always account for some deviation from perfect centralization. The major result of improper centralization is a "smearing" of

WAVEFORMS IN UNFRACTURED ROCKS

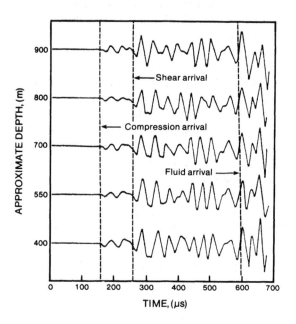

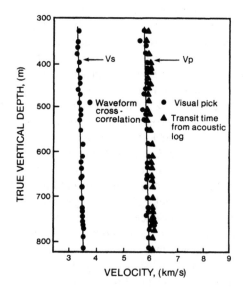

FIGURE 2.4. Example of seismic velocities determined from acoustic fullwave recordings in unfractured granite showing increasing velocity with depth associated with increasing confining pressure.

the acoustic arrivals at the receiver, appearing as a decrease in amplitude and a broadening of the phase spectrum in the detected signal. This occurs because acoustic waves travel along slightly different distances and arrive at slightly different times along different circumferential paths. If the tool is badly decentralized the different phases of the acoustic signal will cancel when averaged at the receiver, making phase arrivals very difficult to measure. The requirement that the received acoustic signal be nearly in phase at all points around the circumference of the borehole even when the logging tool is slightly decentralized places some restrictions on the source frequencies to be used during logging. This amounts to requiring that the source wavelength be somewhat larger than the borehole radius. If suitably long wavelengths are employed, a small decentralization (10 to 20% of the borehole diameter) will not produce circumferential differences in travel paths such that arriving signals are more than a few percent out of phase.

Most acoustic sources in conventional logging equipment excite centerband frequencies in the range from 10 to 30 kHz. These frequencies compare to typical acoustic wavelengths as follows:

| Lithology | $V_p$ (km/s) | $V_s$ (km/s) | P Wavelength (cm) $f_0$=10 kHz | P Wavelength (cm) $f_0$=30 kHz | S Wavelength (cm) $f_0$=10 kHz | S Wavelength (cm) $f_0$=30 kHz |
|---|---|---|---|---|---|---|
| Basalt | 7.0 | 4.2 | 70 | 23 | 42 | 14 |
| Granite | 5.9 | 3.4 | 59 | 20 | 33 | 11 |
| Sandstone | 4.5 | 2.4 | 45 | 15 | 24 | 6 |
| Shale | 2.8 | 1.0 | 28 | 9 | 10 | 3 |

These values indicate that the conventional acoustic logging frequencies usually produce wavelengths slightly greater than borehole diameters for typical exploratory boreholes with diameters ranging from 7.5 to 40 cm. The one important exception is relatively low-velocity shale, where conventional logging frequencies may not provide sufficiently long wavelengths, and acoustic transit time logs might benefit from the use of sources with frequencies below 10 kHz. Acoustic logs are normally obtained with frequencies near 10 or 15 kHz in petroleum reservoirs where lithologies range from porous limestones to shaley sandstones and borehole diameters often are greater than 20 cm. Higher frequencies often are used in mining applications where acoustic velocities of rocks are greater, and borehole diameters as small as 7.5 cm. Thus, the requirements for insentivity to tool tilt and decentralization are satisfied in most applications.

The other important geometrical parameter in acoustic logging is the source to receiver spacing. Three considerations are important in selecting receiver locations: (1) spacings need to be large enough so that the first arriving waves penetrate to a distance beyond the annulus of rock affected by drilling; (2) spacings need to be large enough to achieve far-field wave propagation; and (3) spacings need to be short enough so that waves are not greatly degraded by intrinsic attenuation, outward radiation, and scattering. Superimposed on these considerations is the independent design criterion that the shortest acceptable spacing is the best in terms of optimal vertical resolution. Most consolidated rocks without unusual features such as major fractures (which are especially effective in scattering acoustic energy) permit acoustic waves to be refracted along the borehole with enough energy for effective measurement over distances of at least 10 wavelengths. Natural attenuation of acoustic energy poses a problem for acoustic logging only in the case of extensively fractured or shallow, unconsolidated sediments.

One major concern in source-receiver spacing is the penetration depth of the acoustic waves. The thickness of the annular region in which acoustic logging systems sample rock velocity depends upon the wavelength of the acoustic signal, and the radial velocity distribution caused by alteration of rock properties by stress release and exposure to borehole fluid. Acoustic source frequencies associated with wavelengths equal to or greater than the borehole diameter insure that critically refracted waves traveling along the borehole wall will penetrate to a finite depth. Various numerical experiments have shown that this depth is slightly greater than one wavelength (Baker, 1984). Using this criterion, source to receiver spacings need only be large enough to insure that the far-field wave propagation is set up. Practical experience supported by theoretical consideration indicate that two or more wavelengths will suffice for this purpose.

The situation becomes more complicated when alteration related to stress release, mechanical damage, or softening of clay minerals exposed to drilling fluid affects an annular region of rock around the borehole. In that case, longer source to receiver spacings have been recommended to insure that critically refracted waves represent propagation along the deeper interface between altered and unaltered rock (Koerperich, 1979; Baker, 1984; Baker and Winbow, 1988). The criteria needed to determine how great the source to receiver spacing needs to be to insure measurement of unaltered velocities are obtained from the equations given for refraction in layered media (Grant and West, 1960). These criteria have been evaluated for the borehole by Baker (1984), who found that established separation criteria for first arrivals for layered media agreed with the results of synthetic borehole microseismogram calculations for annular layers around the borehole. In some cases, source to receiver spacing needs to be greater than 10 wavelengths before critically refracted first arrivals penetrate to the unaltered rock beyond the annular region of alteration.

## 2.4    PRACTICAL CONSIDERATIONS IN ACOUSTIC TRANSIT TIME LOGGING

The procedures for obtaining acoustic velocity profiles using conventional acoustic transit time logging equipment appear to be simple and straight forward. The pressure signals detected at the two receivers are transformed into voltages by piezoelectric transducers, the fixed downhole and variable uphole gains are used to amplify the signal so that the first cycle of the detected waveform exceeds the preset threshold, and the detected arrival times are subtracted to determine the interval transit time. This process is illustrated in Figure 2.2. Note that the first part of the receiver waveforms appears similar in the display for each receiver. This happens because both receivers are spaced far enough from the source that the critically refracted waves have assumed their far-field form. The second waveform is delayed with respect to the first, and there are minor changes that could be attributed to attenuation and dispersion. However, visual inspection of the velocity picks displayed at the surface control module for the logging system leaves no doubt that the apparatus is picking the same phase at each receiver. For most consolidated geological formations, natural attenuation is so small that there is a great dynamic range between signal and background noise, and uphole amplification can usually raise the first cycle of the waveform arrival at either receiver far beyond the arrival detection threshold.

In spite of the apparent simplicity of the acoustic transit time picking scheme illustrated in Figure 2.2, a number of things can go wrong. One difficulty arises when the two waveforms from which arrivals are being picked become very different. This occurs when there is a lithology change or significant change in borehole cross section between the near and far receivers which causes a difference in waveform appearance. Because the acoustic arrival detection algorithm is not picking the first energy arrival but the first zero crossing after a threshold has been exceeded, differences in waveform shape can cause errors in velocity measurement even when first arrivals are visually recognizable (Figure 2.5). Such errors associated with lithology contacts are relatively small and not very frequent. The problem becomes much more severe when fracturing, borehole wall washouts, or some other factor greatly attenuates the measured signal. In that case, the first cycles of the detected waveform suddenly fall below the detection threshold, and the wave-picking algorithm triggers on a much later cycle in the waveform (Figure 2.6). This anomaly in the recorded transit times begins as an abrupt shift to very long transit times in uncompensated logging tools when the far detector passes over the fracture first. In borehole compensated acoustic logging probes, this very large transit time is averaged with the correct transit time adjacent to the fracture, producing almost as large a deflection. In both cases, the pen on the strip chart recorder appears to jump abruptly off scale. The transit time picks may then return to an intermediate value as acoustic signals are attenuated at both receivers.

Cycle skips are one of the most important difficulties in acoustic logging, whether for porosity estimation or other purposes. Most modern acoustic logging systems incorporate special processing and filtering methods into data recording. The effects of cycle skipping are removed by rejecting abrupt changes in measured transit times and then smoothing the transit time profile

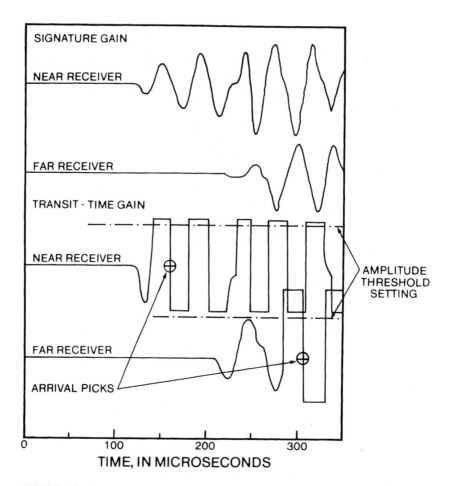

FIGURE 2.5. Illustration of acoustic transit time error caused by change in waveform shape between near and far receiver.

that remains (Figure 2.7). For some applications it may be preferable to preserve a preliminary record of cycle skips in a field copy of the log before removing cycles skips by interpolating data from unaffected intervals because cycle skips provide valuable indications of fracture locations (Keys, 1979).

## 2.5   THEORETICAL PREDICTIONS OF ACOUSTIC VELOCITIES IN POROUS SOLIDS

Although the Wyllie time-average equation is based on empirical data, Gassmann (1951), Geertsma (1957), and Geertsma and Smit (1961) were able to derive a semiquantitative expression for the dependence of compressional velocity on porosity for granular solids. If the mechanical properties of both the

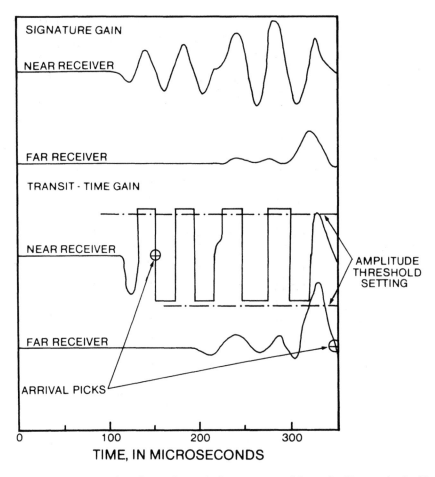

FIGURE 2.6. Illustration of acoustic transit time error caused by cycle skip associated with severe signal attenuation by propagation across a fracture located between near and far receivers.

mineral composing the grains in the solid and the framework composed of those mineral grains are known, the compressional velocity of the porous solid saturated with a fluid is:

$$\rho V_p = \left\{ \frac{1}{C_b} + \frac{4}{3}\mu_b \right\} + \frac{\left(1 - C_s/C_b\right)^2}{\left(1 - \phi - C_s/C_b\right)C_s + \phi C_f} \qquad 2.3$$

where $C_s$ and $C_b$ are the compressibility of the mineral and the framework, $\mu_b$ is the shear modulus of the unsaturated framework, $C_f$ is the compressibility of the fluid, and $\phi$ is the porosity. The introduction of porosity into a granular material is assumed to cause an increase in compressibility. Geertsma (1957),

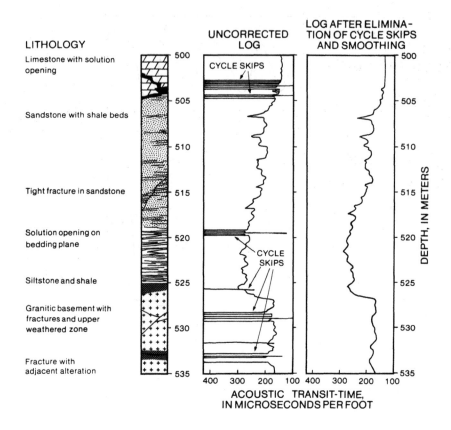

FIGURE 2.7.   Illustration of acoustic transit time log before and after rejection of cycle skips and smoothing.

Brown and Korringa (1975), and Domenico (1977) presented laboratory data indicating a linear relationship between framework compressibility and porosity:

$$C_b = C_s + \alpha\phi \qquad\qquad 2.3A$$

If this equation is substituted into equation 2.3 and the usual binomial expansion used to express the inverse of $V_p$, then Equation 2.3 reduces to a linear relation between transit time ($1/V_p$) and porosity (White, 1983).

The relationship between effective stress and $V_p$ and $V_s$ can be demonstrated using theoretical models of closely packed spheres. These models are reviewed in some detail by White (1983) based on the theory presented by Love (1944) and Timoshenko and Goodier (1951). The expressions for $V_p$ and $V_s$ in simple cubic arrays of spheres preloaded by a pressure $P_0$ are (White, 1983)

$$V_p = \left[ \frac{3E_s^2 P_0}{8(1-V_s)^2} \right]^{1/6} \sqrt{\frac{6}{\pi \rho_s}}$$

*2.4*

$$V_s = \left[ 3(1-v_s)E_s^2 P_0 \right]^{1/6} \sqrt{\frac{3}{(2-v_s)(1+v_s \pi \rho_s)}}$$

where $E_s$, $v_s$, and $\rho_s$ are the Young's modulus, Poisson's ratio, and density for the solid composing the spheres.

The most comprehensive theory for seismic wave propagation in porous and permeable solids is given by Biot (1956a,b, 1962b). The fundamental strain energy density of a fluid-filled porous solid is used to derive a pair of coupled equations for the elastic wave motion in the solid framework and fluid. Although the validity of Biot approach has not been fully establish to the satisfaction of all researchers, several investigators have confirmed the relation between permeability and the existence of a second "slow" compressional wave predicted by this theory (Dutta, 1980; Plona, 1980; Chandler, 1981; Chandler and Johnson, 1981; Feng and Johnson, 1983a,b). The Biot formulation appears to reproduce all of the expected properties of porous media, and is routinely used to calculate the properties of wave propagation in porous solids (Rosenbaum, 1974; Burns, 1988; Schmitt, 1988). A more thorough discussion of the application of the Biot formation to waveform logging will be presented in a subsequent chapter where the effects of porosity and permeability on acoustic waveforms in boreholes and the modeling of acoustic propagation by means of synthetic microseismograms using the Biot (1956a,b) equations is treated in detail.

# 3

# Principals of Seismic Refraction Along Boreholes and Channels

The propagation of acoustic waves in fluid-filled boreholes is a difficult process to describe mathematically. In principal, one can simply define the boundary conditions and solve the appropriate set of wave equations. In practice, the complicated details of the microseismograms generated by numerical integration of the wave equations are often no easier to explain than the characteristics of waveforms obtained in boreholes. The most productive approach to understanding the relationship between waveform character and rock properties has been through a combination of theoretical insight into properties of individual modes of propagation, and direct computation of waveforms for specific conditions. In keeping with this two-sided approach, we shall address both the general process of mode propagation in boreholes and channels in a qualitative sense (this chapter), and the mathematically rigorous computation of waveforms for representative source and boundary conditions (the following chapter).

This chapter presents an overview of wave propagation in the vicinity of fluid-solid interfaces, emphasizing the process whereby acoustic wave energy is trapped in channels and boreholes by internal refraction and constructive interference. The limited computations will treat the plane geometry channel case, where constructive interference effects can be illustrated without the added complications of cylindrical geometry (i.e., Bessel functions are not required). This discussion will set the stage for the following chapter by defining the physical characteristics of the various wave modes, and by providing an understanding of the way in which borehole geometry influences wave excitation. The next chapter will then deal directly with the cylindrical borehole problem in full mathematical detail.

## 3.1 MODELING WAVE PROPAGATION IN CHANNELS AND BOREHOLES

The use of a single source and receiver pair at a given depth in a borehole to measure the seismic velocities of rocks *in situ* appears simple and straightforward. Applications for such acoustic well logs were quickly added to the standard geophysical logging suite. Introductory seismic literature portrays the process of velocity logging in wells as either another porosity log, or as a means of augmenting surface seismic soundings. In the latter case, the comparison between borehole velocity logs and surface seismic reflection surveys is viewed as a trade-off in which the direct access to depth information is weighed against the confusing amount of detail interjected by the much finer vertical resolution of the well log. However, detailed investigation of the seismic refraction problem, in which a seismic signal is generated along the borehole axis and a transmitted signal recorded some distance away along the borehole axis, demonstrates that this wave propagation problem is a great deal more complicated than simple transmission through a small volume of rock surrounding the borehole.

The most difficult aspect of the borehole wave propagation problem is the large (in fact dominant) contribution of refracted wave energy to the measured response at an acoustic receiver in a conventional logging probe. The complexity of the mathematical problem is not very relevant when first arrivals are the only subject of interest. The mathematical details become much more relevant in the case where the geophysicist attempts to extract additional information on rock properties by investigating the complete waveform recorded at one or more receiver stations. This chapter presents a physical discussion of the waveguide refraction properties of the borehole as they affect the character and interpretation of waveforms recorded by modern acoustic waveform logging systems.

Although various waveform logging systems use somewhat different frequencies, bandwidths, source to receiver spacings, and down-hole filtering systems, several decades of experimentation have produced acoustic logging systems with the configuration illustrated in Figure 2.1. The empirical design of such systems was carried out many years before the appropriate wave propagation problem could be solved on modern high-speed computers. However, experience with system performance resulted in selection of source frequencies within the range of 10 to 30 kHz, yielding seismic wavelengths ranging from 10 to 100 cm, or from slightly less than borehole diameter to several borehole diameters. Source to receiver spacings range from one to several seismic wavelengths, and occasionally somewhat longer. As will be shown below, these values are not an arbitrary result of tool fabrication with available components; the tool characteristics illustrated in Figure 1.2 represent an optimum for velocity measurement using refracted compressional waves. In the widening field of applications for acoustic full waveform logs, other measurements likely will require application of different frequency ranges or different spacings. A sound understanding of principles of borehole wave propagation will be important in the design of equipment for such applications.

## 3.2 REFRACTION AND REFLECTION AT A PLANE INTERFACE

The design of acoustic logging equipment (i.e., location of receivers some distance away from the source along the borehole axis) makes this form of seismic measurement sensitive to the wave energy refracted along the borehole. Because the rock surrounding the borehole has one or both seismic velocities ($V_p$ and $V_s$) greater than that of the fluid filling the borehole, the fluid-filled borehole acts as a waveguide, the efficiency of which is determined by the specific borehole geometry, velocity contrasts, and frequencies excited by the logging source.

The reflection and transmission coefficients for unattenuated, plane parallel compressional waves in a fluid incident on a plane, infinite interface between fluid and solid are given in Figure 3.1. The coefficients are independent of frequency and wavelength because there is no characteristic length scale involved; the entire problem is determined by the velocity contrasts between the fluid and the solid, and by the angle of incidence of the wave fronts. In the case illustrated, both the compressional and shear velocities of the solid are greater than the speed of sound in the fluid. Unlike the more general case of an interface between two elastic solids, reflected shear waves are not generated in the fluid. Very little transmitted shear is produced for low angles of incidence, but transmitted shear becomes important as the angle of incidence increases.

### 3.2.1 Critical Angle of Refraction and Complete Reflection of Acoustic Energy

One of the most important characteristics of the curves illustrated in Figure 3.1 is the critical angle of incidence for incoming wave fronts, beyond which no energy is transmitted into the solid. The critical angle decreases for increasing velocity contrasts, so that the angle of incidence beyond which energy is not transmitted into the solid is determined by the shear velocity (i.e., the velocity closest to the acoustic velocity of the fluid). The physical consequence of such a limit is that energy incident at the interface at angles of incidence beyond the critical angle for shear transmission is fully trapped within the fluid. In more general cases when some transmission occurs we will refer to "leaking" or "leaky" modes as wave propagation in which seismic energy traveling along the borehole-fluid interface decreases with distance because wave energy is being radiated outward. In a common situation where both shear and compressional velocities in the rock around the borehole exceed the speed of sound in the borehole fluid ($V_p > V_s > V_f$), the conditions for fully trapped wave propagation will depend upon the shear velocity of the rock. In other situations typical of "soft" clay mineral rich or shallow, poorly consolidated sediments, shear velocity is less than the acoustic velocity of the borehole fluid ($V_p > V_f > V_s$).

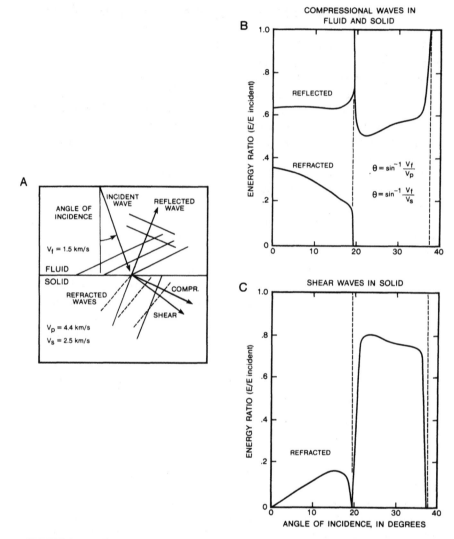

FIGURE 3.1.  Reflection and transmission coefficients for plane waves in a fluid impinging on a solid surface.

In that case, the compressional velocity in the rock determines the character of trapped energy propagation along the borehole, and characteristics of the full waveform logs generated become very different from the "hard" rock case.

### 3.2.2   Interface Waves Generated by a Point Source

In the plane interface case illustrated in Figure 3.1, the characteristics of the problem are completely determined by the velocity contrast at the interface. In a slightly more realistic case, the characteristic of the acoustic energy source

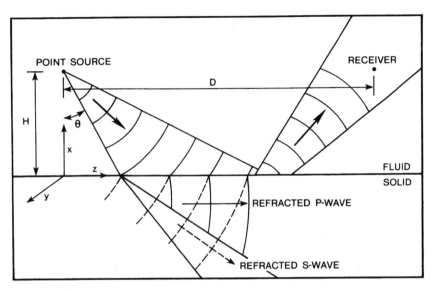

FIGURE 3.2.   Acoustic wave propagation between a point source located a distance $H$ over a solid half-space and a receiver located a distance $D$ away.

can be introduced as in Figure 3.2. Here the problem is expanded to include a point source located at a distance $H$ above the interface and a point receiver located a similar distance above the interface a distance $D$ away. Although this appears to be a modest increase in complexity in the problem, the mathematical complications introduced are significant. The finite separations of source and receiver, and the finite distance of both above the interface introduce characteristic length scales. The distance may be chosen large enough such that far-field approximations can be applied. The distance introduces a fundamental ratio, $H/L$ (where $L$ is the acoustic wavelength in the fluid) such that the problem in Figure 3.2 reduces to the plane wave incidence problem only in the limit $H/L \to \infty$.

The formal solution to the general problem in Figure 3.2 involves Fourier decomposition of the point source and solution of the wave equation by enforcement of boundary conditions at the interface. The solution is given as a double integral over frequency and wavenumber (parallel to the interface) space (Ewing et al., 1957; Roever et al., 1959):

$$P = P_0 + P_1$$

$$P_1 = C_0 \int_{-\infty}^{\infty} X(\omega) e^{-i\omega t} d\omega \int_{-\infty}^{\infty} \frac{N(x, y, \kappa, \omega)}{D(\kappa, \omega)} e^{i\kappa z} d\kappa \qquad \qquad 3.1$$

Where $P_0(x,y,z,t)$ is the source function, and $C_0$ is a constant depending on the magnitude of $P_0$. In this expression $P(x,y,z,t)$ is the pressure in the borehole fluid, which consists of the source pressure field, $P_0$ and the reflected pressure field $P_1$; $X(\omega)$ is a source spectrum; $N(x,y,\kappa,\omega)$ and $D(\kappa,\omega)$ are determined by the boundary conditions, and $\kappa$ and $\omega$ are the wavenumber and frequency of the transformed variables (vertical coordinate, $z$, and time, $t$). The formal evaluation of this integral introduces several factors which will be important in the fluid-filled borehole case. Following the notation of Ewing et al. (1957), the integral 3.1 can be divided into contributions from singularities in the denominator (zeros of the function $D(\kappa,\omega)$) and portions of the complex integration path deformed to travel around branch cuts (Peterson, 1974; Tsang and Rader, 1979). Standard procedure for evaluation of Equation 3.1 is to compute wave excitation at a given frequency by integration with respect to wavenumber, followed by Fourier inversion in time. This allows application of several different source spectra for a given wavenumber integration.

The singularities in Equation 3.1 represent wave modes excited on the interface between solid and fluid. These are the familiar Rayleigh and Stoneley interface modes (Rayleigh, 1885; Stoneley, 1949). The excitation of each can be calculated separately by residue theory, or from the composite waveform generation by direct integration of Equation 3.1. In this case, neither exhibit much dispersion; that is, phase velocity is nearly independent of frequency. The dominant source of attenuation for these modes is the geometrical spreading which results as the waves propagate on the two-dimensional interface away from the point underneath the source. A full discussion of the general solution to Equation 3.1 is given by Roever et al. (1959). One important result from that analysis is the introduction of a slight amount of dispersion into the Rayleigh mode. Roever et al. (1959) designate this as the "pseudo-Rayleigh" mode to account for this difference. The dispersion is caused by the small amount of attenuation associated with radiation of wave energy into the overlying fluid which results from the disturbance produced by the propagation of the surface waves. The "pseudo Rayleigh" mode described by Roever et al. (1959) is distinctly different from the borehole pseudo-Rayleigh mode described by recent authors.

## 3.3    CRITICAL REFRACTION AND THE GENERATION OF HEAD WAVES

Propagation of wave energy along the fluid-solid interface is of primary importance for the geophysicist interested in wave propagation along a fluid-filled borehole. The character and travel time of early energy arrivals in the overlying fluid in Figure 3.2 is determined by the propagation of waves in the solid just below the interface, traveling at velocities (compressional and shear) greater than the acoustic velocities in the fluid. Seismologists have observed experimentally that such waves, denoted as head waves in the literature (Grant and West, 1960), produce a measurable signal in the overlying fluid half-space. These waves are shown to determine the first arrivals in waveforms generated by the formal integration of Equation 3.1. In the years before such integration was made possible by modern high-speed computers, the head wave component to the solution to Equation 3.1 could be represented as integrals around branch cuts in the complex plane (Ewing et al., 1957; Brekhovskikh, 1960). Such deformations of the integration path are formalities of the mathematics. They arise because of the paired sets of incoming and outgoing wave solutions to the governing wave equation; branch cuts are necessary to insure that boundary conditions are properly enforced. The branch cut integrations effectively represent the interface head waves because they incorporate all the physics related to radiation losses transmitted away from the interface.

Branch cuts and branch cut integrals turn out to be important for the calculation of head waves in boreholes, and must be discussed in some detail. The branch cuts arise because the mathematics of wave propagation require the use of complex variables, and complex square roots produce multiple defined functions. The branch cuts represent lines set up on the computational plane that cannot be crossed, so that we (or the computer) always know which value of the complex square root to use. The issue is important because the two roots represent two possible solutions to the governing differential equation. Some of these solutions are rejected because they represent incoming radiation — and cannot be allowed according to the radiation boundary conditions.

Several authors have used complex algebra to show how head waves can be calculated by integrating along the branch lines on either side of the cut. At first this appears to be a mathematical accident. We shall find that the association of branch cut integrals with head waves represents an important physical process. The head waves represent the summed contribution of wave fronts that are almost (but not quite) critically refracted along the borehole. These waves are losing energy through a small amount of transmission away from the borehole. The cumulative effect of energy loss at different frequencies acts to determine the characteristics of the head waves we measure. The difference in the computed Fourier amplitude of the waves on either side of the branch cut therefore represents the effect of these radiation losses in the complex arithmetic. So in spite of the formidable-looking algebra, the computations are doing what geophysicists usually do — keeping track of transmission and reflection at the fluid-solid interface.

## 3.4  ACOUSTIC PROPAGATION IN THE FLUID-FILLED CHANNEL WAVEGUIDE

The point source and receiver problem in Figure 3.2 appears to contain all of the mathematical aspects of borehole waveform theory, except for one fundamental difference. A single interface does not allow for internal trapping of wave energy within the fluid by means of complete internal refraction. The geometric properties of the fluid-filled borehole allow for both partial and complete trapping of wave energy, greatly enhancing the amplitude of acoustic signals detected at positions along the borehole axis. This condition takes two forms: (1) removal of geometric spreading (but not dispersion) by constraining waves to travel in one rather than two dimensions and (2) allowing for partial or complete internal refraction of acoustic and seismic waves. The first of these is self evident and needs little explanation. The second turns out to be the single most important factor in understanding the properties of seismic waves in boreholes.

### 3.4.1   The Primary Role of Geometry — Constructive Interference and Trapped Modes

The limited size of the fluid-filled borehole or channel in comparison with the infinite width of the fluid half-space in Figure 3.2 introduces the possibility for constructive interference. In contrast, the curved surface of the cylindrical interface in a fluid-filled borehole adds very little that is new to the fluid channel problem, except in the limit of infinite wavelength and low frequencies. The laterally enclosed fluid geometry introduces the possibility of a number of additional trapped modes caused by refraction and constructive interference. This process is illustrated in Figure 3.3. Consider a finite channel of width $2a$ separating two semi-infinite elastic half-spaces. There will be an infinite series of possible constructively interfering wave fronts where reflections at the channel wall exactly reinforce incoming wave fronts:

$$nL\cos\varphi = \alpha; \text{ Constructive interface condition}$$
$$L = \text{acoustic wavelength}$$

*3.2*

The angle of incidence of each wave front satisfying the constructive interference relation is related to the width of the channel, $2a$. As the channel becomes wider, an increasing number of the possible constructive interferences can occur with angle of incidence beyond critical. For these modes only, all energy is trapped within the waveguide. The rest of the infinite set of wave fronts in Equation 3.2 lose energy on each reflection because a certain fraction of the incident energy is transmitted and lost. We shall refer to these partially transmitted modes as "PT" modes in future discussions.

Equation 3.2 describes the infinite series (or modes) of wave fronts of various angles of incidence that constructively interfere for a given frequency and channel width. In predicting wave propagation, we consider a source characterized by a certain frequency spectrum and radiation pattern. Such a source excites a finite number of modes The criteria for constructive interference and complete internal reflection combine to define a cutoff frequency for each of the series of internal reflection modes defined by Equation 3.2 that are excited by the source. These cutoffs define those frequencies below which there is no possibility of complete internal reflection. Note that Equation 3.2 gives the successively higher cutoff frequencies as simple multiples of the

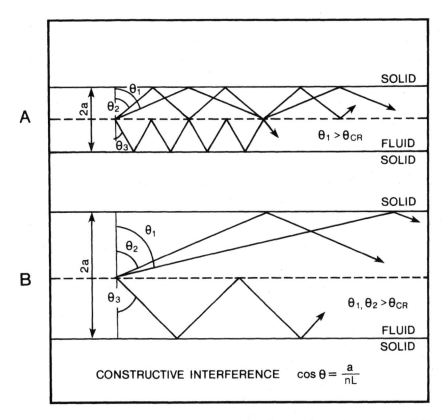

FIGURE 3.3.  Diagram illustrating the constructive interference of acoustic waves in a fluid between two elastic half-spaces: (A) small channel width allowing one complete internal reflection, and (B) larger channel width allowing two modes with complete internal reflection.

lowest cutoff. For finite source frequency bandwidth, only a finite number of fully trapped modes can be generated.

Equation 3.2 was derived on the assumption of very small wavelengths such that acoustic waves in the fluid-filled channel are simply reflected from the channel walls (the classic ray approximation). If we allow finite wavelengths, the seismic velocities of the elastic half-spaces on either side of the channel enter into the physics of the wave reflections. The most important complication is the splitting of the constructively interfering modes into two families related to the compressional and shear velocities of the solid. The proportions with which source energy is partitioned into shear and compressional modes depends upon the velocity contrasts involved. For velocities typical of "hard" rocks, more of the energy goes into shear modes, whereas almost all energy goes into compressional modes for "soft" rocks. But in theory, both sets of modes exist for all cases, and each has its own ascending series of cutoff frequencies below which undamped propagation is not possible.

These general comments about wave propagation along a fluid-filled channel embedded between two elastic half-spaces indicate the fate of energy radiating outward from a source placed in the fluid. With the exception of those wave fronts satisfying conditions for constructive interference, almost all wave energy is radiated outwards with a bit of distortion to the otherwise spherically symmetric wave fronts imposed by the velocity contrasts at the channel edges. Those few wave fronts for which constructive interferences are possible are preferentially propagated along the fluid channel. The highly distorted spectrum of this limited component of the original source spectrum will determine the character of the pressure arrivals at a detector some distance away in the channel. In the far-field limit, the waveform recorded in the fluid will be completely determined by the contributions from leaking and fully trapped modes.

### 3.4.2    The Plane Channel Analog to the Fluid-Filled Borehole

The general comments given in the preceding paragraph imply that there are some analogies between the cylindrical column of fluid filling a borehole and a channel of fluid between two elastic half-spaces. The physical connection between the two models is provided by the dominance of constructive fluid interference effects and the splitting of such internally reflected modes into sets of compressional and shear modes. The difference in geometry, and especially the use of cylindrical coordinates of the borehole case introduce some quantitative differences. However, the imposition of symmetry requirements on the fluid slab problem increases the correspondence between wave propagation in the two different cases. The equations for the fluid channel problem involve simple exponentials rather than linear combinations of Bessel functions. The fluid channel analogy therefore gives an excellent qualitative introduction to the physical mechanisms of mode trapping and constructive interference that characterize waveform logs.

The fluid channel analog to the fluid-filled borehole is illustrated in Figure 3.4. The channel half width, $a$, is analogous to the borehole radius. In addition, we impose symmetry about the central plane of the channel in order to restrict consideration to those solutions that would be symmetrical about the borehole axis. The equations are reduced to two dimensions in space ($x$ and $z$) by assuming all displacements are constant and uniform with respect to the third ($y$) direction. Throughout the rest of this chapter we will assume that acoustic sources and detectors are located on the central plane of the channel. Under those assumptions, we assume solutions to the wave equation in the channel of the form:

$$\text{Radial displacement} = u_r = \frac{\partial \phi_f}{\partial r} \qquad\qquad 3.3$$

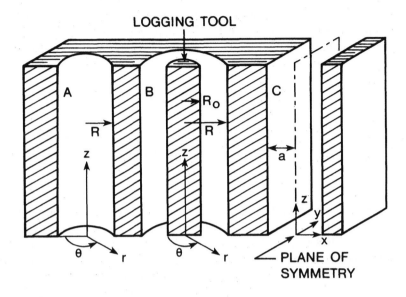

FIGURE 3.4.  The plane geometry analog to the fluid-filled borehole: (A) fluid-filled borehole; (B) fluid-filled annulus between rigid logging tool and borehole wall; and (C) plane channel with central plane of symmetry such that channel halfwidth is analogous to annulus width or borehole radius.

$$\phi_f = D \cosh\left(m_1 x\right) e^{i(\kappa z - \omega t)}; \quad m_1^{\,2} = \kappa^2 - \frac{\omega^2}{V_f^{\,2}}$$

and in the elastic half space of the form:

$$u_r = \frac{\partial \phi_s}{\partial r} - \frac{\partial \psi_s}{\partial z}$$

$$\phi_s = A \cosh\left(m_1 x\right) e^{i\left(m_2 x + \kappa z - \omega t\right)}; \quad m_2^{\,2} = \kappa^2 - \frac{\omega^2}{V_p^{\,2}}$$

$$\psi_f = B \cosh\left(m_1 x\right) e^{i\left(m_3 x + \kappa z - \omega t\right)}; \quad m_3^{\,2} = \kappa^2 - \frac{\omega^2}{V_s^{\,2}}$$

3.4

Radiation conditions (no incoming waves) force us to reject inward propagating solutions in the elastic solid, and boundary conditions at the solid-fluid interface provide constraints among the constants $A$ and $B$ in the solutions. The usual procedure for evaluation of the boundary condition as coupled linear equations and matching with source amplitude provides a solution of the form given by Equation 3.1:

$$P(x,z,t) = P_0 + P_1$$

$$P_1(x,z,\omega) = \int_{-\infty}^{\infty} \frac{N(\kappa,\omega)}{D(\kappa,\omega)} e^{i\kappa z} d\kappa \qquad 3.5$$

The pressure response function $N/D$ contains a number of singularities that correspond to the cases of complete internal reflection of constructively interfering modes. If the integral in Equation 3.1 were evaluated by conventional complex variable methods, the contribution to the measured pressure response could be determined by the usual residue techniques. The complex integration procedure, now rarely used, in the evaluation of the integral in Equation 3.1 is outlined in Appendix 3.1.

### 3.4.3    Fully Trapped Symmetric Modes and Head Waves

Evaluation of $D$ in Equation 3.1 gives the following form to the denominator (Paillet and White, 1982):

$$D = \left(\kappa^2 - m_3^2\right) m_2 \omega^2 \frac{\rho_f}{\rho_s} \cosh(m_1 a) + \left\{ \left(\kappa^2 + m_3^2\right)^2 - 4 m_2 m_3 \kappa^2 \right\} m_1 V_s^2 \sinh(m_1 a)$$

$$3.6$$

The singularities in the complex wavenumber plane for given value of frequency, $\omega$, are given by the zeros in this relation. We find one such zero corresponding to the Stoneley mode for all frequencies. In addition, we find successively more zeros as frequency is increased past the cutoff frequencies for higher modes. For the two families of critically refracted compressional and shear waves, the cutoff frequencies for each mode increase as the channel width is decreased. A finite number of these zeros exist when Equation 3.6 is evaluated for values of $\omega$ above cutoff for the lowest mode. Two infinite series of additional solutions to Equation 3.6 corresponding to attenuated or leaky compressional and shear modes, can be found by allowing for complex values of the wavenumber $\kappa$. There are many more zeros of Equation 3.6 that correspond to those compressional and shear modes with cutoff frequencies greater than the frequency range of interest at which the integral in Equation 3.1 is being evaluated.

The two infinite series of modes represent those internally reflected waves that constructively interfere during propagation. They have been split into two separate families because the "reflection" occurs by means of refraction of both shear and compressional modes at the fluid-solid interface. The Stoneley mode (Stoneley, 1949) represents a separate mode that exists at all frequencies for the fluid-filled channel. A typical set of dispersion curves showing group and phase velocity for a water filled channel 20 cm wide ($a = 10$ cm) and separating two elastic half spaces with seismic velocities typical of sandstone is shown in Figure 3.5. At the low frequency cutoff for each shear mode, mode phase velocity is equal to shear velocity in the elastic solid, but falls asymptotically towards acoustic velocity ($V_f$) in the fluid as frequency increases. Mode group velocity starts at shear velocity at the cutoff frequency, drops towards a minimum below acoustic velocity in the fluid at higher frequencies, and then eventually increases towards $V_f$ again. The Stoneley mode propagates at phase and group velocities slightly less than acoustic velocity in the fluid, but Stoneley velocity falls towards zero in the low frequency limit. This is the one significant difference between mode propagation in fluid-filled channels and boreholes. In the borehole case, Stoneley phase velocity maintains a finite value in the low frequency limit (Paillet and White, 1982; White, 1983).

### 3.4.4    Critically Refracted Head Waves and First Arrivals

The amplitude and characteristics of waveforms for each of the modes associated with zeros in Equation 3.6 can be evaluated by residue theory in wavenumber space and Fourier synthesis in time. How are the compressional and shear head waves calculated? These are the first arrivals that one would use to measure compressional and shear velocity. Although it is not obvious without extensive mathematical manipulation and transformation of variables, Peterson (1974) and Rosenbaum (1974) demonstrated that the computation of the head wave contribution can be calculated by branch cut integrations analogous to those for the single plane interface given by Ewing et al. (1957).

According to the early literature, waveforms measured in boreholes and fluid-filled channels were considered composites of waveforms from a finite series of modes, each associated with a specific residue calculation, and the shear and compressional head waves, each given by a Fourier branch cut integral over the complete source spectrum. The finite series of modes, each associated with a zero of the dispersion relation of Equation 3.5 include the Stoneley mode, and two sets of compressional and shear modes. This apparently natural partitioning of the waveform energy still left some loose ends: (1) those "leaky" modes for which the singularly lies off of the real wave number axis in 3.5, and (2) the Rayleigh surface wave mode known to propagate along the surface of elastic bodies (Rayleigh, 1885; Roever et al., 1959). How do "leaky" and partially transmitted (PT) modes and Rayleigh waves figure into the characteristics of the waves we measure? The leaky and PT modes will be

A    PHASE VELOCITY

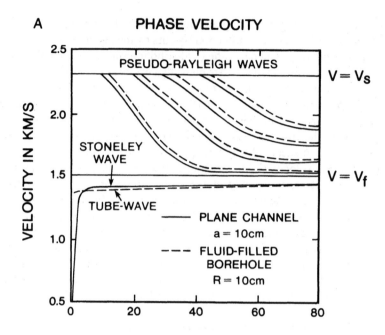

B    GROUP VELOCITY

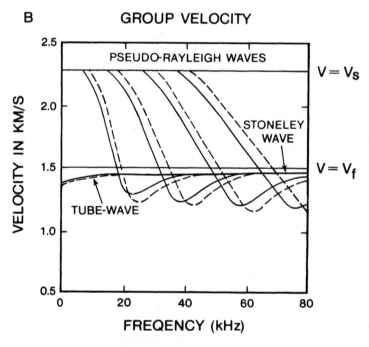

FIGURE 3.5. Pseudo-Rayleigh (first four modes) and Stoneley wave phase and group velocities for a 10 cm channel halfwidth compared to pseudo-Rayleigh and tube-wave mode phase and group velocities in a 10 cm radius borehole.

discussed in this section of Chapter 3, and the Rayleigh wave question will be addressed in a subsequent section.

### 3.4.5  Guided Modes and Head Waves

After a great deal of computation, it turns out that the PT modes completely determine compressional and shear head waves in the far field. This happens because one can show that the contribution to the integrand in Equation 3.1 in the vicinity of the PT modes completely dominates the branch cut integral when source to receiver separations become large. The numerical domination stems from the oscillatory character of the term $e^{i\kappa z}$ in the wavenumber integration as the source to receiver separation becomes large. In that case, adjacent cycles of the oscillatory integrand cancel each other during the integration, except where the factor multiplying $e^{i\kappa z}$ in the integrand varies rapidly. At large enough separations, the integration becomes dominated by those specific frequencies where PT modes affect the integrand. In the far-field limit, the PT modes alone determine the head waves. It was shown rigorously by Rosenbaum (1960) and later by Haddon (1987, 1989) that a sufficient number of transformations of variable could be used to express head waves as an infinite series summed over the residues for all modes below cutoff rather than a branch cut integration. All of this sounds like useless mathematical trivia, except that an understanding of how leaky modes contribute to the head wave gives a useful insight into how a specific acoustic energy source excites measurable head waves. It will turn out that the relationship between modes and head wave amplitude can be used to explain how certain frequency ranges and source to receiver separations optimize velocity measurements in boreholes.

To begin with, we have to generalize the characteristics of partially transmitted (PT) modes and their relationship to the shear and compressional modes. Going back to the idea of constructively interfering waves in the fluid, we can think of two infinite sets of modes for each frequency. One represents shear refraction in the solid, and the other compressional refraction. At very low frequencies, none of the waves are critically refracted so that all modes are classified as PT modes. At sufficiently large frequencies, one or more of the constructive interferences for each mode will become completely reflected; these are the compressional and shear normal modes. In the formal mathematics of the problem, the poles associated with each mode are shown to move about on the complex plane. The presence of branch cuts accounts for the contribution of partially transmitted (PT) modes to the integrand in Equation 3.1. The modes track from a lower Riemann sheet through the branch point at mode cutoff, and out along the real wavenumber axis as frequencies increase (Paillet and Cheng, 1986; Kurkjian, 1985; Haddon, 1987). This process is illustrated schematically in Figure 3.6.

The effect of leaky modes on the head wave integration occurs because a pole located on a lower Riemann sheet can cause a large contribution to the

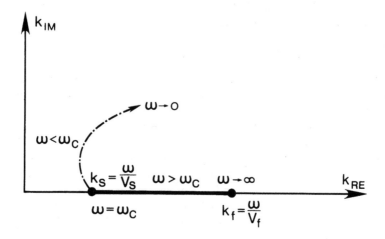

FIGURE 3.6.  Trajectory of shear normal mode pole as frequency decreases through mode cutoff; the pole drops down onto a lower Riemann sheet where the computation cannot yield a root of the dispersion Equation 3.6 without taking the negative root of the expression $m_3 = (\kappa^2 - \omega^2/V_s)^{1/2}$.

integration over those narrow frequency ranges where the pole lies close to the branch point. This process is illustrated in Figure 3.7, where the integrand in Equation 3.1 for a typical fluid-filled channel is shown for frequencies just above and below the cutoff for one of the shear normal modes. Above cutoff, the integrand shows the expected excursions of a singular function. This is the guided wave contribution associated with complete internal refraction within the borehole waveguide. Below cutoff, the integrand shows a sharp "spike", with an apparent discontinuity in slope at the branch point. The near discontinuity results from the difference in amplitude of the contribution from the leaky mode pole on either side of the branch point. In the limit of large source to receiver separations, this near discontinuity in the integrand dominates the integration of the head wave with the composite waveform.

The practical importance of these complicated mathematical arguments is that the head wave response measured in fluid-filled channels and boreholes will consist of a number of spectral peaks, each associated with the short frequency interval where one particular mode lies just below mode cutoff. Optimal excitation of head waves will result when the acoustic source closely corresponds to one of these peaks. The excitation of multiple peaks by a broad

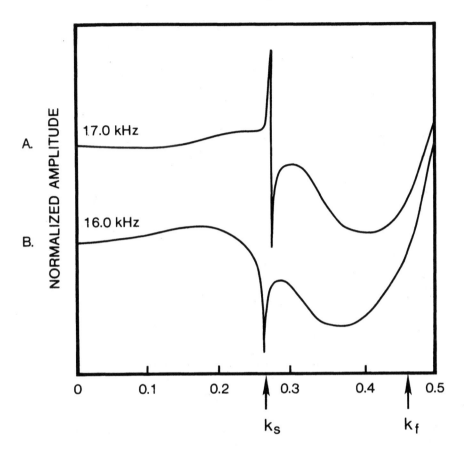

FIGURE 3.7. Comparison of the pressure response function for the channel when (A) frequency is above cutoff and shear normal mode lies on the upper Riemann sheet, and (B) frequency is below cutoff and shear normal mode lies on the lower sheet near the branch point.

band source will cause borehole propagation to distort the original source signature, complicating any attempts at deconvolution of the observed waveforms (Figure 3.8) (Paillet, 1981a; Paillet and Cheng, 1986). This distortion can also be described as the interference between multiple modes of refracted head waves excited by a broad band source. In either case, it appears that the design of acoustic logging equipment to excite the borehole at frequencies corresponding to the cutoff for the first compressional and shear modes would be the most efficient method for producing usable head waves. Acoustic

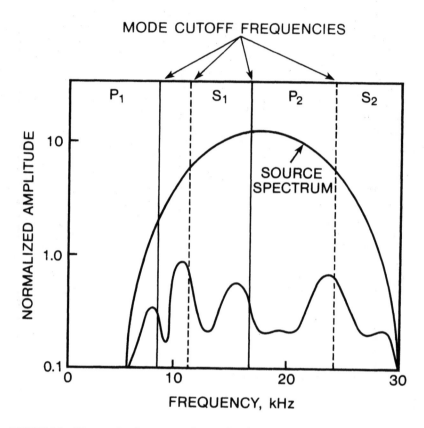

FIGURE 3.8. Diagram showing computed spectral peaks in Fourier coefficients for head waves associated with mode cutoff.

logging equipment has been designed by trial and error over decades of use to correspond to just that frequency range in typical boreholes.

### 3.4.6 The Missing Rayleigh Mode

One of the more confusing differences between wave propagation on the interface between a fluid and elastic half space (Figure 3.2) and in the fluid-filled channel between two elastic half-spaces is the disappearance of the familiar Rayleigh pole from the period Equation 3.6 for the channel. What has become of the Rayleigh pole? The simplest way to see the relationship between the Rayleigh pole and the normal modes in the borehole is to consider the period equation for the fluid-filled channel in the limit of vanishing density of the borehole fluid. In that case, the period Equation 3.6 reduces to the product of two terms, one of which yields the familiar Rayleigh pole:

$$\left(\kappa^2 + m_3{}^2\right) - 4m_2 m_3 \kappa^2 = 0 \ \left(\text{Rayleigh mode in the limit } \frac{\rho_f}{\rho_s} \to 0\right) \quad 3.7$$

The other factor in the second term in Equation 3.6 represents the poles corresponding to constructively interfering waves from effectively rigid walls. The elastic half-spaces on either side of the fluid-filled channel appear rigid to the fluid because of the infinite contrast in density. The limit of vanishing fluid density indicates that the normal mode singularities found for the critically refracted waves are a complicated hybridization of the Rayleigh and reflected modes. The coupling between these two modes as the density of the fluid in the channel is increased is illustrated in Figure 3.9. The separate dispersion curves for the Rayleigh mode at the surface of the elastic half-spaces and the reflected modes within the channel transform into a single set of curves that are distorted adjacent to the Rayleigh velocity for small values of the density ratio. At greater values of the density ratio, the dispersion curves become the familiar normal modes.

The disappearance of the Rayleigh pole for finite value of fluid density substantiates that the Rayleigh singularity has not been overlooked in the analysis. The Rayleigh pole does not appear in the denominator of the integrand in Equation 3.1 because the coupling of fluid in the channel to wave motion at the surface of the elastic half-spaces completely alters the mathematics of the problem. The Rayleigh surface wave becomes incorporated into the shear normal mode. For this reason, the shear normal modes in waveform logging often are denoted as pseudo-Rayleigh waves. However, the distortion of the dispersion curve in Figure 3.9 for finite values of the density ratio indicate the great difference between Rayleigh surface waves and shear normal modes. The terminology issue is even more complicated, because the term pseudo-Rayleigh was originally introduced to denote the relatively minor dispersion and attenuation introduced in the case where a fluid half-space overlies an elastic solid (Roever et al., 1959). The great changes in the properties of wave propagation introduced by the hybridization of Rayleigh and reflected fluid modes underscores the complications introduced by the possibility of internal reflections in the fluid.

Rayleigh surface waves modified slightly by the curvature of the borehole wall are shown to exist in cylindrical cavities (Biot, 1952; Bostrom and Burden, 1982). These modes hybridize with internal reflections in the borehole when a fluid of finite density is introduced into the cavity in a manner closely analogous to the hybridization of modes given by Equation 3.6, producing mode dispersion curves similar to those illustrated in Figure 3.9.

## 3.5 MODE SUMMARY

The properties of the various modes present in the fluid channel waveguide appear complicated even without cylindrical geometry. However, the actual modes can be divided into three sets: (1) the slightly dispersive Stoneley mode associated with a single pole; (2) the strongly dispersive shear normal modes

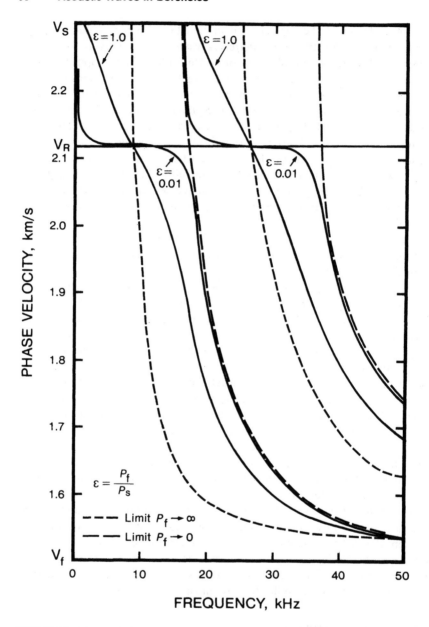

FIGURE 3.9. Phase velocities for normal modes in the limit of vanishing fluid density illustrate the uncoupling of internal reflections and Rayleigh mode in the limit $\rho_f/\rho_s \to 0$.

(or pseudo-Rayleigh modes in cylindrical boreholes) associated with an infinite series of poles, only some of which correspond to fully trapped, nonleaking modes; and (3) a similar series of dispersive compressional normal modes associated with another infinite set of poles. For the purposes of calculation, the two infinite sets of shear and compressional modes are divided into subsets of the finite set above their cutoff at a given frequency, and all of the rest. The mathematical theory demonstrates that the infinite set of modes below cutoff determine the properties of the first arrivals (the head waves). In theory, the compressional and shear head waves can be represented by an infinite series summed over the residues at all of these poles (Rosenbaum, 1960; Haddon, 1987). The properties of these modes are summarized in Table 3.1. In the table we distinguish between two kinds of attenuated modes. We use "PL" to denote compressional modes above cutoff in which shear radiation induces losses. We use "PT" to denote "partially transmitted" or "partially refracted" compressional and shear modes in which frequencies lie below cutoff, and losses occur because waves are partially transmitted at the rock-fluid interface.

Numerical investigations demonstrate that head wave contributions are determined by modes just below cutoff. These contributions introduce sharp spectral peaks at those particular frequencies where individual modes are slightly below cutoff. At lower frequencies the modes are highly damped. At the same time, energy arguments show that mode amplitude must vanish at the cutoff frequency. That is, wave motion in the elastic solid is attached to wave fronts that are normal to the channel walls when refracted modes correspond to the cutoff frequencies. Such motion would require infinite energy to excite at even the smallest amplitude. Investigation of acoustic logging equipment developed through decades of practical experience indicates that such equipment has been designed so that source frequency bands span the cutoff frequency for the lowest set of compressional and shear normal modes under typical logging conditions (i.e., those imposed by seismic velocities and borehole diameter). If lower frequencies were used, compressional and shear head waves would not be generated. If higher frequencies or wider frequency bands are used, multiple spectral peaks caused by multiple mode cutoffs produce complicated and distorted first arrivals. At the same time, the complete head-wave theory for channels and boreholes demonstrate the many ways in which seismic waves in boreholes differ from simple body waves in uniform elastic solids.

## 3.6  MODE CONTENT, CHANNEL WIDTH, AND FREQUENCY SCALING

An interesting by-product of the guided wave mode analysis is the number and varieties of modes present over the imposed frequency band. Because the frequencies and amplitudes of modes depend upon the width of the channel, there is a certain interchangeability between channel width and frequency. That is, one can scale source frequency bands in such a way as to produce similar waveforms for different channel widths. This relationship between frequency and channel width will be demonstrated numerically for the cylindrical borehole in the next chapter. Both the channel and borehole case are similar in that

**Table 3.1**
**Mode Summary — Plane Channels, Symmetric Modes in Borehole**

| Mode | Type | Radiation Trapping | Approximate Range Group Velocities ($V_m$=Mode Velocity) |
|---|---|---|---|
| Fast, Hard Formations — $V_p > V_s > V_f$ | | | |
| Tube wave (Stoneley) | Stoneley interface | Fully trapped, slight dispersion | $0.75V_f < V_m < V_f$ |
| Pseudo-Rayleigh (shear normal mode) | Fully refracted shear | Fully trapped, dispersive | $0.8V_f < V_m < V_s$ |
| PT shear | Partially transmitted refracted shear | Leakage due to incomplete refraction; this mode governs shear head wave. | $V_s{}^{\text{a}}$ |
| PL (leaky "P") | Fully refracted compressional | Leakage due to loss by conversion to shear during refraction | $V_m < V_p$ |
| PT compressional | Partially transmitted compressional | Leakage due to incomplete refraction; this mode governs compressional head wave. | $V_p{}^{\text{a}}$ |
| Slow, Soft Formations — $V_p > V_f > V_s$ | | | |
| Tube wave (Stoneley) | Stoneley interface slight dispersion | Fully trapped | $0.7V_f < V_m < V_s{}^{\text{b}}$ |
| Leaky tube wave | Stoneley interface | Partially trapped—some shear radiation losses | $V_s < V_m < {}^{**}0.7V_f$ |
| PL (leaky "P") | Fully refracted compressional | Leakage due to loss by conversion to shear during refraction | $0.7V_f < V_m < V_p{}^{\text{b}}$ |
| PT compressional | Partially transmitted | Leakage due to incomplete refracted compressional refraction | $V_p{}^{\text{a}}$ |

[a] Although these modes track on a lower Riemann sheet, their contribution to the synthetic seismogram occurs through the branch point such that propagation velocities always correspond to $V_p$ and $V_s$.

[b] $0.7V_f$ limit based on density ratio $\rho_s/\rho_f = 2.0$; limit slightly different for other density ratios.

waveform properties are primarily controlled by the mode content of the waveform. One of the most important practical applications for synthetic borehole microseismograms is the use of calculations to define frequency bands and other characteristics associated with the optimum excitation of the wave modes of interest.

# Appendix 3.1

## Complex Contour Integration Evaluation of Wavenumber Transform for Synthetic Microseismograms

Although direct numerical evaluation of transform integrals such as that in Equation 3.1 is now the preferred method of synthetic microseismogram calculation, complex contour integration methods can provide considerable insight into the properties of waveforms. Complex contour integration and residue theory also provide important conceptual tools in the illustration of waveform properties and in the isolation of the contribution of specific modes of wave propagation to the composite waveform. The use of different Fourier integral conventions by different authors has produced some confusion about the way in which singularities contribute to the Fourier inversion in wavenumber and frequency. This confusion has been compounded further by the use of various asymptotic expansions and transformations of variable. Here we present an outline of the complex contour integration procedure using the Fourier conventions given in Equation 3.1. These same conventions are used throughout the synthetic microseismogram computations given in Chapters 4 to 8.

The wavenumber inversion integral for synthetic microseismogram computations is given in the form:

$$P(x,z,\omega) = \int_{-\infty}^{\infty} \frac{N(\kappa,\omega)}{D(\kappa,\omega)} e^{i\kappa z} d\kappa \qquad A3.1$$

The usual procedure for evaluating the double integral in Equation 3.1 is to first perform the wavenumber integration in Equation A3.1. We will focus attention on the wavenumber integration as the first step in generating a series of Fourier coefficients representing the pressure response of the borehole to excitation by a point source applied at a single harmonic frequency, $\omega$. The series of values obtained for a number of different frequencies may then be convolved with a given source frequency spectrum to produce the desired microseismogram.

Although we focus on the wavenumber integration, the sign convention in the frequency inversion is important because the combination of signs determines the direction of wave propagation. With the combination of negative sign for frequency and positive sign for wavenumber we insure that wave propagation is in the upward (positive z) direction for positive real wavenumbers and frequencies. Equally important, positive imaginary wavenumbers are associated with damped wave propagation because they indicate exponential decrease with distance for upward propagating waves. The identities of the

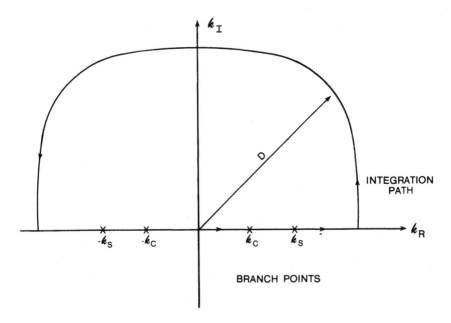

FIGURE A3.1.    Integration path and return contour in the complex wavenumber plane used to formulate inversion of wavenumber Fourier integral using residue theory.

various signs corresponding to damped motion can be worked out for any choice of Fourier convention, but we prefer the version given in Equation 3.1 because positive real wavenumbers are always associated with propagation in the positive (upward) direction, and negative wavenumbers with negative (downward) propagation.

The complex contour integration evaluation of the integral in Equation 3.1 is based on the principle that integrals of analytic functions of a complex variable along a closed contour can be related to the sum of the residues at each of the poles enclosed by the contour (McLachlan, 1955). The integrand in Equation A3.1 is found to have the required analytic properties in the region defined by excluding the branch cuts associated with complex square roots. We can include the complex integration indicated in Equation A3.1 in a closed contour as illustrated in Figure A3.1. Here we have closed the contour by taking the "return" path at a radius D from the origin, and then take the limit $D \to \infty$. We have chosen a return path above the real wavenumber axis so that the imaginary part of the wavenumber along the integration path will always be very large and positive. This produces a factor in the integrand which vanishes in the limit $D \to \infty$, and the contribution to the total integral along the return path vanishes.

Before we can relate the value of the Fourier integral to the sum of the residues enclosed by the integration path, we have to account for departures from the return contour associated with the branch cuts. We follow Ewing et

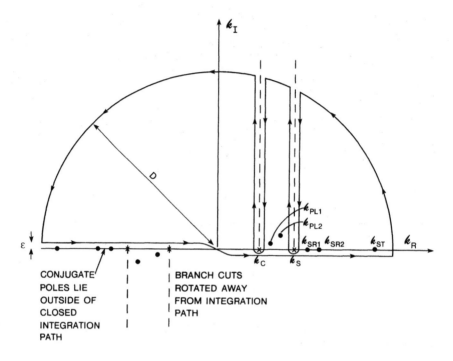

FIGURE A3.2.   Integration path in Figure A3.1 modified to account for radiation conditions and detours of return path required to avoid crossing branch cuts in the complex wavenumber plane.

al. (1957) and Peterson (1974) in selecting branch cuts as illustrated in Figure A3.2. At first it is not clear that the branch cuts need to be accounted for at all. We cannot tell whether the integration path passes under or above the branch points. If the branch cuts could be arranged below the real axis, it seems that we might avoid them altogether. However, consideration of radiation conditions shows that the integration path passes under the real axis for positive real wavenumbers, and above the real axis for negative real wavenumbers (Kurkjian, 1983, 1985).

The configuration of branch cuts illustrated in Figure A3.2 indicates that only one pair of compressional and shear branch cuts affect the integration. If we identify the integrals required to deform the contours along the two branch cuts as

$$I_p = \int_{\kappa_p - \varepsilon}^{\kappa_p - \varepsilon + i\infty} \frac{N}{D} e^{i\kappa z} \, d\kappa - \int_{\kappa_p + \varepsilon}^{\kappa_p + \varepsilon + i\infty} \frac{N}{D} e^{i\kappa z} \, d\kappa$$

$$I_s = \int_{\kappa_s - \varepsilon}^{\kappa_s - \varepsilon + i\infty} \frac{N}{D} e^{i\varepsilon z} \, d\kappa - \int_{\kappa_s + \varepsilon}^{\kappa_s + \varepsilon + i\infty} \frac{N}{D} e^{i\kappa z} \, d\kappa$$

A3.2

where $\kappa_p = \omega/V_p$, $\kappa_s = \omega/V_s$, and the integrals are evaluated in the limit $\varepsilon \to 0$. Then the integral in Equation A3.1 can be evaluated as:

$$\int_{-\infty}^{\infty} \frac{N}{D} e^{i\kappa z} d\kappa = -\left\{ I_p + I_s + 2\pi i \Sigma R_i \right\} \qquad A3.3$$

where $R_i$ are the residues associated with the compressional normal mode (PL wave), shear normal mode (pseudo-Rayleigh wave), and Stoneley (tube wave) singularities. The other compressional and shear singularities (partially transmitted) still contribute to the integral in Equation A3.3 through their effect on the integrals $I_p$ and $I_s$.

In the case of a single plane interface, Ewing et al. (1957) divide the far-field wave into the source, modes (Stoneley and Rayleigh waves represented by contributions from residues) and head waves (represented by branch cut integrals $I_p$ and $I_s$). We can carry the analysis a few steps further in the case of waves trapped in a fluid-filled borehole or channel. When waves are trapped in a channel, we find that all contribution to $I_p$ and $I_s$ is associated with the specific frequency values where the individual mode singularities "pass" through the branch point. The contribution is naturally divided into two parts (those above and those below mode cutoff) because energy conservation requires that wave amplitude vanish when mode singularities approach the branch point (Paillet and White, 1982). The coalescence of the mode pole with the branch point corresponds to mode propagation at shear or compressional velocity, where energy arguments require that mode amplitudes vanish. Each mode contributes to the branch cut integral over the narrow frequency range where that mode is approaching the branch point as frequency increases toward mode cutoff, and mode amplitude has not become very small. Mode amplitude again increases with increasing frequency, when the pole becomes a singularity on the upper sheet moves away from the branch point, but this contribution is given by the residues in Equation A3.3. Kurkjian (1985) uses direct numerical evaluation of the shear branch cut integral to distinguish the shear head wave from the pseudo-Rayleigh modes, finding that the method works, but that great accuracy is required in the computation to insure that numerical noise does not dominate the calculated microseimogram.

The complex contour integration evaluation of the Fourier inversion integral in Equation A3.1 indicates how the composite microseismogram is naturally partitioned into various wave modes. Shear normal modes (or pseudo-Rayleigh waves) and the Stoneley mode (tube wave) are given by residues at poles encircled by the complex integration contour. It also turns out that the leaky compressional mode or PL mode is analogous to the shear normal mode, except that the PL mode is damped by the loss of shear wave energy to the formation. The PL mode still lies of the upper Riemann sheet in one of the two quadrants encircled by the integration contour, and still can be accounted for by residue theory. The compressional and shear head waves are given by the branch cut integrals. In the far field, the contribution to these integrals is completely dominated by the compressional and shear mode poles when they approach cut off frequencies from below. This contribution dominates the integrals $I_p$ and $I_s$ because the modes below cutoff also lie on a lower Riemann sheet. These modes produce large differences in the integrand on either side of the branch cut. In contrast, the modes above cut off lie on the upper sheet, and therefore within the region of analytic behavior of the integrand. These upper sheet modes cannot contribute to differences in the integrand no matter how large they become.

This discussion completes the story of complex contour evaluation of the integral in Equation A3.1. All of the further numerical complications one sees in the literature result from additional attempts to evaluate the integrals $I_p$ and $I_s$. Several authors (Rosenbaum, 1960; Peterson, 1974; Tsang and Rader, 1979; Haddon, 1987, 1989) use transformations of variables and asymptotic series expansions for various arguments in the integrands to determine additional properties of wave modes. These investigations confirm that the head waves can be formally represented as a series summed over the residues of all of the PT modes (Table 3.1) on the lower Riemann sheet after suitable transformations of variables and analytic extensions of the integrand.

# 4

# Synthetic Microseismograms in Fluid-Filled Boreholes

In the previous chapter we reviewed the theory of acoustic wave propagation in channels and cylindrical boreholes in a general way. The review pointed out the various differences between acoustic propagation along an interface and acoustic propagation within a cylinder or channel where partial or complete internal refraction of wave energy can produce guided waves that dominate far-field measurements. The discussion emphasized the identification of individual modes and relied in a general way on the methods of the theory of complex variable. In previous decades, these methods were employed in conjunction with asymptotic representations of Bessel functions to calculate the waveforms produced by an acoustic source in a borehole. Most of this formalism has been swept away by modern high-speed computers. Integrals such as Equation 3.1 can simply be integrated over a given frequency band to provide a synthetic microseismogram output for prescribed borehole geometry and source to receiver separation. The singularities encountered in evaluation of the integral are avoided by the simple expedient of including a small amount of attenuation in the wave equation (see discussion of "convergence factors" in complex integration; Morse and Feshback, 1953). One can assume a small but finite value for the complex part of the wave velocity ($V_I = \omega_I/\kappa$) in the governing equations in the elastic solid. This small attenuation can then either be identified with the expected attenuation in real rocks (i.e., assume typical values of $Q_p$ and $Q_s$) or be removed from the solution by multiplication with the appropriate increasing exponential function afterwards. This chapter reviews the formal evaluation of the Fourier integral solution to the wave propagation problem for the fluid-filled borehole using this direct, numerical approach.

## 4.1 SYNTHETIC MICROSEISMOGRAMS — MODELING BOREHOLE, FLUID, AND LOGGING TOOL

Acoustic propagation in a fluid-filled borehole is modeled according to the diagram in Figure 4.1. A cylindrical logging tool of diameter $2R_0$ is centralized in a fluid-filled cylindrical cavity of diameter $2R$ and is surrounded by an infinite

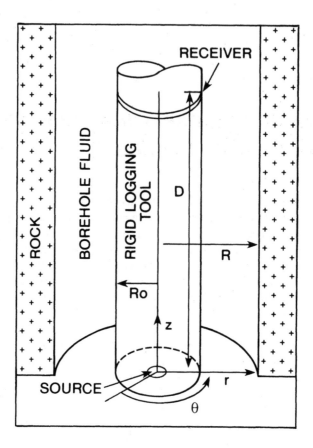

FIGURE 4.1. Diagram showing logging tool of radius $R_0$ centralized in fluid-filled borehole of radius $R$.

elastic solid. The model is specified by the density and acoustic velocity of the borehole fluid, $\rho_f$ and $V_f$, and the density and seismic velocities of the surrounding rock, $\rho_s$, $V_p$ and $V_s$. Waves in the borehole are applied along the surface of the logging tool ($z = 0$; $r = R_0$), and fluid pressure is evaluated at the surface of the logging tool at a point some distance away ($z = D$; $r = R_0$). The most general solution to the wave equations in the fluid-filled borehole can be expanded in cylindrical harmonics. In many examples of wave generation from point and line sources, the coefficients in the expansion in cylindrical harmonics are found to decrease with increasing azimuthal periodicity. The largest coefficient in the series might therefore be expected to be the lowest order term in the series. Acoustic logging probes have taken advantage of this expected result with transducer configurations designed to excite only the fundamental, azimuthally symmetric mode. All higher mode energy that might be excited by the source is filtered from the data by averaging the measured pressure response around the circumference of each receiver. For this reason, most of the early numerical models of full waveform logs have assumed such symmetry (Biot, 1952; White, 1962; Roever et al., 1974; Schoenberg et al., 1981; Cheng and Toksöz, 1981; White and Zechman, 1968). We first derive the equations for the case of azimuthal symmetry which is clearly the most useful numerical model for geophysical applications. We then return to the original equations to develop more complicated solutions for those lower nonsymmetric modes which prove to be of interest.

If cylindrical symmetry is assumed, motion consists of "conical waves" traveling along the borehole, and displacements are restricted to the $z$-$r$ plane (Morse, 1939; Jacobi, 1949; Biot, 1952; Somers, 1953; Zemanek and Rudnick, 1961). Using cylindrical coordinates, vector displacements $\mathbf{u}$ in the fluid and solid can be represented by a combination of a scalar potential, $\phi$, and vector potential, $\xi$, such that

$$\mathbf{u} = \left(u_r, u_\theta, u_z\right) = \nabla\phi + \nabla \times \xi \qquad 4.1$$

In the axisymmetric case, only the azimuthal component, $\psi$, of the vector potential is nonzero. The radial displacement may then be represented as

$$u_r = \frac{\partial\phi}{\partial r} - \frac{\partial\psi}{\partial z} \qquad 4.2$$

where $\phi$ is the scalar potential and $\psi$ is the azimuthal (and only nonzero) component of the vector potential. Similarly, the radial and tangential stresses are

$$\sigma_{rr} = \rho\left(\frac{\nu}{1-\nu}\right)\frac{\partial^2\phi}{\partial t^2} + 2\mu\left(\frac{\partial^2\phi}{\partial r^2} - \frac{\partial^2\psi}{\partial z^2}\right) \qquad 4.3$$

and

$$\sigma_{rz} = \rho\frac{\partial^2\psi}{\partial t^2} + 2\mu\left(\frac{\partial^2\phi}{\partial r\partial z} - \frac{\partial^2\psi}{\partial z^2}\right) \qquad 4.4$$

where $\nu$ is Poisson's ratio, $\mu$ the shear modulus, and $\rho$ the density. The equations of motion are

$$\frac{\partial^2\phi}{\partial r^2} + \frac{1}{r}\frac{\partial\phi}{\partial z^2} = \frac{1}{V_p^2}\frac{\partial^2\phi}{\partial t^2} \qquad 4.5$$

and

$$\frac{\partial^2 \psi}{\partial r^2} + \frac{1}{r}\frac{\partial \psi}{\partial r} - \frac{\psi}{r^2} + \frac{\partial^2 \psi}{\partial z^2} = \frac{1}{V_s^2}\frac{\partial^2 \psi}{\partial t^2} \qquad\qquad 4.6$$

where $V_p$ and $V_s$ are the compressional and shear-wave velocities. The general solutions for conical waves are

$$\phi = \left[A_1 K_0(m_2 r) + A_2 I_0(m_2 r)\right]e^{i(\kappa z - \omega t)} \qquad\qquad 4.7$$

and

$$\psi = \left[B_1 K_1(m_3 r) + B_2 I_1(m_3 r)\right]e^{i(\kappa z - \omega t)} \qquad\qquad 4.8$$

where $A_1, A_2, B_1, B_2$ are constants, $\kappa$ is the wavenumber in the $z$-direction, $K_1$ and $I_1$ are the modified Bessel functions of the ith order, and $m_2$ and $m_3$ are the wavenumbers in the radial direction, given by

$$m_2{}^2 = \kappa^2 - \frac{\omega^2}{V_p{}^2}$$

and

$$m_3{}^2 = \kappa^2 - \frac{\omega^2}{V_s{}^2}$$

$$4.9$$

The boundary conditions (1) at $r = R$, the boundary between the rock formation and the borehole fluid, and (2) at $r = R_0$, the boundary between the tool and the fluid, are that the radial stress and displacement are continuous and the tangential stress is zero. In addition, the radiation condition requires that the waves vanish at infinity and that stresses and displacements remain finite at $r = 0$, the center of the borehole.

In the fluid, the pressure $P_f$ and displacement $u_f$ are given by

$$P_f = -\rho_f \frac{\partial^2 \phi_f}{\partial t^2} \qquad\qquad 4.10$$

and

$$u_r = \frac{\partial \phi_f}{\partial_r}$$ 4.11

The scalar potential can be written as

$$\phi = \left[ D_1 I_0 \left( m_1 r \right) + D_2 K_0 \left( m_1 r \right) \right] e^{i \left( \kappa z - \omega t \right)}$$ 4.12

where

$$m_1^2 = \kappa^2 - \frac{\omega^2}{V_f^2}$$

Using the relationships (Abramowitz and Stegun, 1964)

$$\frac{d I_0(y)}{dy} = I_1(y)$$

and

$$\frac{d K_0(y)}{dy} = -K_1(y)$$

the pressure and radial displacement in the fluid can be expressed as

$$P_f = \rho_f \omega^2 \left[ D_1 I_0 \left( m_1 r \right) + D_2 K_0 \left( m_1 r \right) \right] e^{i \left( \kappa z - \omega t \right)}$$ 4.13

$$u_r = m_1 \left[ D_1 I_1 \left( m_1 r \right) - D_2 K_1 \left( m_1 r \right) \right] e^{i \left( \kappa z - \omega t \right)}$$ 4.14

In the formation around the borehole, the only solutions allowed by the radiation condition are of the $K_0$ and $K_1$ type, thus

$$\phi_s = A_1 K_0 \left( m_2 r \right) e^{i \left( \kappa z - \omega t \right)}$$

and 4.15

$$\psi_s = B_1 K_1 \left( m_3 r \right) e^{i \left( \kappa z - \omega t \right)}$$

The radial displacement and stress and the tangential stress in the solid can thus be written as

$$u_r = \left[ A_1 m_2 K'_0 (m_2 r) - i\kappa_1 (m_3 r) \right] e^{i(\kappa z - \omega t)}$$

$$\sigma_{rr} = \left\{ -\rho_s \left( \frac{V_p^2 - 2V_s^2}{V_p^2} \right) \omega^2 A_1 K_0 (m_2 r) \right. \tag{4.16}$$

$$\left. + 2\rho_s V_s^2 \left[ A_1 m_2 K''_0 (m_2 r) - i\kappa B_1 m_3 K_1 (m_3 r) \right] \right\} e^{i(\kappa z - \omega t)} \tag{4.17}$$

and

$$\sigma_{rz} = \left\{ -\rho_s \omega^2 B_1 K_1 (m_3 r) \right\} + 2\rho_s V_s^2 \left[ i\kappa A_1 m_2 K'_0 (m_2 r) + \kappa^2 B_1 K_1 (m_3 r) \right] \right\} e^{i(\kappa z - \omega t)}$$

$$\tag{4.18}$$

The boundary conditions at $r = R$ require $\sigma_{rz} = 0$, so

$$\kappa^2 B_1 K_1 (m_3 R) = \frac{2\omega^2 V_s^2}{\omega^2 - 2\kappa^2 V_s^2} \left\{ i\kappa A_1 m_2 K'_0 (m_2 R) \right\} \tag{4.19}$$

This, together with the relationship (Abramowitz and Stegun, 1964)

$$\frac{d^2 K_0(y)}{dy^2} = K_0(y) + \frac{1}{y} K_1(y)$$

gives, after some algebraic manipulations, the radial displacement and stress in the solid at $r = R$:

$$u_r = \frac{\omega^2}{2\kappa^2 V_s^2 - \omega^2} \left[ A_1 m_2 K_1 (m_2 R) \right] e^{i(\kappa z - \omega t)}$$

and $\tag{4.20}$

$$\sigma_{rr} = \rho_s \omega^2 A_1 \left\{ \frac{2\kappa^2 V_s^2 - \omega^2}{\omega^2} K_0 (m_2 R) \right.$$

$$+\frac{2V_s^2 m_2 m_3 K_1(m_2 R)}{\omega^2 - 2\kappa^2 V_s^2}\left[\frac{1}{m_3 R} + \frac{2\kappa^2 V_s^2 K_0(m_3 R)}{\omega^2}\frac{}{K_1(m_3 R)}\right]\Bigg\} e^{i(\kappa z - \omega t)} \qquad 4.21$$

At the surface of the logging tool, after applying the boundary condition at $r = R_0$ and the condition that the potentials be finite at $r = 0$, we have

$$u_r^{(tool)} = \frac{\omega^2}{2\kappa^2 V_{st}^2 - \omega^2}\left[A_4 m_4 I_1(m_4 R_0)\right] e^{i(\kappa z - \omega t)}$$

and $\qquad\qquad\qquad\qquad\qquad\qquad\qquad\qquad\qquad\qquad\qquad\qquad 4.22$

$$\sigma_{rr}^{(tool)} = \rho_s \omega^2 A_3 \left\{\frac{2\kappa^2 V_{st}^2 - \omega^2}{\omega^2} I_0(m_4 R_0)\right.$$

$$+\frac{2V_{st}^2 m_4 m_5 I_1(m_4 R_0)}{\omega^2 - 2\kappa^2 V_{st}^2}\left[\frac{1}{m_5 R_0} + \frac{2\kappa^2 V_{st}^2 I_0(m_3 R_0)}{\omega^2}\frac{}{I_1(m_3 R_0)}\right]\Bigg\} e^{i(\kappa z - \omega t)} \qquad 4.23$$

where $A_4$ is the amplitude of the scalar potential in the solid tool, and

$$m_4^2 = \kappa^2 - \frac{\omega^2}{V_{pt}^2}$$

and $\qquad\qquad\qquad\qquad\qquad\qquad\qquad\qquad\qquad\qquad\qquad\qquad 4.24$

$$m_5^2 = \kappa^2 - \frac{\omega^2}{V_{st}^2}$$

$V_{pt}$ and $V_{st}$ are the compressional and shear velocities of the steel logging tool.

## 4.2 BOUNDARY CONDITIONS AND DERIVATION OF THE PERIOD EQUATIONS FOR TRAPPED MODES

We begin by considering conditions allowing wave propagation when there is no borehole forcing ($P_0 = 0$ in Equation 3.1). This approach gives solutions for which trapped modes can propagate without energy losses. Applying the continuity of radial displacement and stress at the two boundaries, we get four coupled equations:

$$\frac{\omega^2}{2\kappa^2 V_s^2 - \omega^2} m_2 K_1(m_2 R) A_1 - m_1 I_1(m_1 R) D_1 + m_1 K_1(m_1 R) D_2 = 0$$

$$\frac{\rho_s}{\rho_f} \left\{ \frac{2\kappa^2 V_s^2 - \omega^2}{\omega^2} K_0(m_2 R) + \frac{2 V_s^2 m_2 m_3 K_1(m_2 R)}{\omega^2 - 2\kappa^2 V_s^2} \left[ \frac{1}{m_3 R} + \frac{2\kappa^2 V_s^2 K_0(m_3 R)}{\omega^2 \, K_1(m_3 R)} \right] \right\} A_1 \qquad 4.25$$

$$+ I_0(m_1 R) D_1 + K_0(m_1 R) D_2 = 0 \qquad 4.26$$

$$m_1 I_1(m_1 R_0) D_1 + m_1 K_1(m_1 R_0) D_2 + \frac{\omega^2}{2\kappa^2 V_{st}^2 - \omega^2} m_4 K_1(m_4 R_0) A_4 = 0$$

and

$$I_0(m_1 R_0) D_1 + K_0(m_1 R_0) D_2 + \frac{\rho_t}{\rho_f} \left\{ \frac{2\kappa^2 V_{st}^2 - \omega^2}{\omega^2} I_0(m_4 R_0) \right. \qquad 4.27$$

$$\left. + \frac{2 V_{st}^2 m_4 m_5 I_1(m_4 R_0)}{\omega^2 - 2\kappa^2 V_{st}^2} \left[ \frac{1}{m_5 R_0} + \frac{2\kappa^2 V_{st}^2 I_0(m_5 R_0)}{\omega^2 \, I_1(m_5 R_0)} \right] \right\} A_1 = 0 \qquad 4.28$$

These equations have the form

$$\begin{pmatrix} \theta_{11} & \theta_{12} & \theta_{13} & 0 \\ \theta_{21} & \theta_{22} & \theta_{23} & 0 \\ 0 & \theta_{32} & \theta_{33} & \theta_{34} \\ 0 & \theta_{42} & \theta_{43} & \theta_{44} \end{pmatrix} \begin{pmatrix} A_1 \\ D_1 \\ D_2 \\ A_4 \end{pmatrix} = 0 \qquad 4.29$$

For this system of equations to have a nontrivial solution, the determinant of the matrix of coefficients $\theta_{ij}$ must vanish. This gives us the period equation for the phase velocity of the pseudo-Rayleigh and Stoneley waves as a function of the wave number $\kappa$. The period equation is

$$\text{Det}\left(\theta_{ij}\right) = 0$$

or

$$\left(\theta_{11}\theta_{22} - \theta_{21}\theta_{12}\right)\left(\theta_{33}\theta_{44} - \theta_{34}\theta_{43}\right) - \left(\theta_{11}\theta_{23} - \theta_{13}\theta_{21}\right)\left(\theta_{32}\theta_{44} - \theta_{34}\theta_{42}\right) = 0$$

*4.30*

The zeros of this period equation give the phase velocity of the guided waves identified for the channel propagation problem in the previous chapter. We refer to the cylindrical counterpart of the Stoneley mode as the tube wave because of the cylindrical geometry in the borehole problem, and because the "hoop" stress effects of the cylindrical wall become prominent at large wavelengths. The uncoupling of the Rayleigh wall waves and refracted fluid waves in the limit $\rho_f/\rho_s \to 0$ was discussed in the previous chapter. The only modification to the theory for the cylindrical waves is that the Rayleigh surface wave remains slightly dispersive even in the limit of $\rho_f/\rho_s \to 0$ as described for surface waves in empty boreholes by Biot (1952) and Bostrom and Burden (1982). The hybridized modes combining the nature of reflected waves in the fluid and critical refraction in the borehole wall are known as pseudo-Rayleigh waves in the case where shear velocity in the rock is larger than the acoustic velocity in the fluid. In some of the literature on borehole waveforms these modes are called shear normal modes to emphasize the significant differences between this mode of propagation and the Rayleigh mode on plane interfaces, and because Roever et al. (1959) have already used the term pseudo-Rayleigh to define a different mode of wave propagation. The term normal mode is also used in the seismic literature to denote channel waves propagating along low velocity wave guides (Rosenbaum, 1960; Aki and Richards, 1980). When shear velocity in the solid falls below the acoustic velocity in the fluid, trapped modes are dominated by critical refraction of compressional waves in the borehole wall, and the modes are known as leaky compressional, or PL modes (Phinney, 1961).

The assignment of a finite set of velocities, $V_{pt}$ and $V_{st}$, to the logging tool in Figure 4.2 adds very little to the properties of the solution as long as both tool velocities are much greater than the acoustic velocity in the borehole fluid. Calculations by Cheng and Toksöz (1980, 1981) indicate that there is little difference between solutions with large but finite values for $V_{pt}$ and $V_{st}$, and solutions with a rigid logging tool in the borehole as given by White and Zechman (1968) and Paillet and White (1982). In the case where the steel tool is considered to be a rigid boundary, the boundary conditions are modified to require no displacement at the surface of the tool ($R_0$) and the fourth condition is dropped entirely. Then the period equation becomes

$$(\theta_{11}\theta_{22} - \theta_{21}\theta_{12})\theta_{33} - (\theta_{11}\theta_{23} - \theta_{13}\theta_{21})\theta_{32} = 0 \qquad \textit{4.31}$$

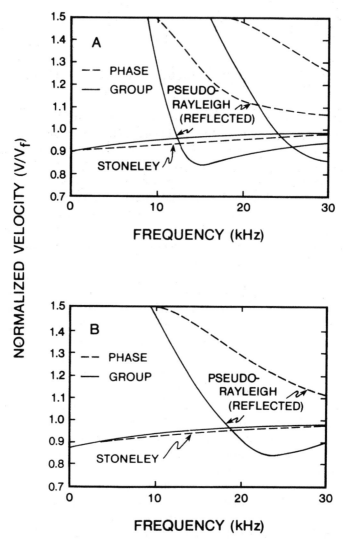

**FIGURE 4.2.** Phase and group velocities for pseudo-Rayleigh and tube-wave modes for (A) fluid-filled borehole and (B) fluid-filled annulus between steel logging tool and borehole wall; $V_p = 4.57$ km/s, $V_s = 2.74$ km/s, $V_f = 1.83$ km/s, $R = 10.2$ cm, $R_0 = 4.6$ cm, $\rho_f = 1.2$ g/cm$^3$, $\rho_s = 2.3$ g/cm$^3$.

For the case of a fluid-filled borehole without a tool (Biot, 1952) the condition that the potentials remain finite at $r = 0$ eliminates the $K_0$ and $K_1$ solutions in the fluid, so that the period equation simplifies to

$$\theta_{11}\theta_{22} - \theta_{21}\theta_{12} = 0 \qquad\qquad 4.32$$

In Equation 4.32 the borehole diameter only appears in the dimensionless product $m_1R$. In that case, the modes and head waves (the properties of the latter being determined by the extension of modes below cutoff frequencies) are rigorously scaled according to the ratio of borehole diameter to wavelength.

The boundary conditions given in Equations 4.24 through 4.27 represent conditions for natural oscillations in the fluid-filled borehole. Existence of such natural oscillations require vanishing of the determinant of the coefficients in Equation 4.29, in which case the amplitude of the oscillation is undetermined. In computing synthetic microseismograms for specific logging applications, borehole forcing at the surface of the logging tool body, or along the borehole axis is assumed. The forcing terms show up as a column matrix replacing the zero on the right side of Equation 4.29. If complex seismic velocities are used to introduce damping into the system of equations, zeros of the determinant do not occur for real values of frequency and wavenumber, and the system of equations can be solved by conventional matrix manipulations.

In most of the synthetic seismogram calculations presented in the following chapters, the system of Equation 4.29 with point source forcing will be solved to give an analytical expression for solution coefficients $D_1$ and $D_2$ (or $D_1$ alone if no logging tool is present) for the displacement in the borehole fluid as a function of wavenumber, $\kappa$, for a specific value of frequency, $\omega$. Numerical integration of these expressions across the real wavenumber axis then yields the set of Fourier coefficients for the borehole forcing. The final microseismogram is computed by convolving the set of stored Fourier coefficients with a specific source frequency spectrum characterized by a centerband frequency, $\omega_0$.

## 4.3   PROPERTIES OF THE PSEUDO-RAYLEIGH AND TUBE WAVE MODES

Before considering the direct computation of the microseismogram for a specific source excitation function, the period equation can be used to investigate the properties of the fully trapped tube wave and pseudo-Rayleigh or shear normal modes. We can give an example by using the borehole parameters given by Biot (1952) for a typical sandstone. In that case, the shear velocity is greater than the acoustic velocity in the fluid, so that pseudo-Rayleigh waves are generated. Solutions for the phase velocity of modes existing at frequencies less than 30 kHz are illustrated in Figure 4.2. The tube wave is shown to exist at all frequencies in this range, with a small increase in velocity from a finite value near $0.90\ V_f$ in the low frequency limit to more than $0.95\ V_f$. Two other solutions to the period equation are shown to exist at successively greater values of frequency for the fluid-filled borehole without logging tool (Figure 4.2A). The lower limits of these two modes represent the cutoff frequencies for the pseudo-Rayleigh modes. In each case, the cutoff frequencies indicate the frequency below which the conditions for constructive interference of wavefronts do not permit complete internal refraction.

Phase velocities of wave propagation often do not give a useful representation of measurable wave arrivals. That is, the measured arrival of wave energy at a location in the borehole is the energy of a finite pulse composed of the superposition of many Fourier modes propagating at different phase velocities. The velocity of the pulse is often more closely represented by the group velocity defined when the period equation is used as a relation between frequency and wavenumber:

$$\omega = \omega(\kappa)$$

$$V_g(\omega_0) = \left.\frac{d\omega(\kappa)}{d\kappa}\right|_{\omega = \omega_0} \qquad 4.33$$

where $\omega(\kappa)$ is taken to represent the set of frequency values satisfying the period equations 4.29, 4.30, or 4.31 for specified values of wavenumber $\kappa$. Group velocities for the tube wave and pseudo-Rayleigh modes are also plotted in Figure 4.2. The small change in tube wave phase velocity with frequency yields a group velocity that is very close to phase velocity. However, this is not the case for the pseudo-Rayleigh modes. Although pseudo-Rayleigh phase velocities drop smoothly from the shear velocity of the elastic solid surrounding the borehole to the acoustic velocity of the borehole fluid, pseudo-Rayleigh group velocity decreases rapidly from $V_s$ at the cutoff frequency to less than $V_f$, and then increases slowly back towards $V_f$ as frequency increases. The presence of a minimum in group velocity produces a phenomenon known as the Airy phase, in which mode energy falls away abruptly after arrival times associated with minimum group velocity. A similar set of modes related to critical refraction of compressional waves in the elastic solid exists, but these modes are not shown in Figure 4.2. The compressional modes are associated with another set of zeros of the period equation, but the preferential excitement of shear refractions when $V_s$ is greater than $V_f$ makes these zeros much more difficult to find by conventional numerical procedures.

The formal solutions to the period equation confirm the dominant effect of channel width or borehole diameter described in the preceding chapter (Cheng and Toksöz, 1981; Paillet and White, 1982; Kurkjian, 1983, 1986). The existence of undamped pseudo-Rayleigh modes is related to the critical refraction at the solid-fluid interface coupled with constructive interference of acoustic waves in the borehole fluid. The series of incidence angles required for internal reflection is determined by borehole diameter, making hole size or width of the annulus between logging tool and borehole wall one of the most important controls on the appearance and quality of acoustic waveforms. In general, more of the internally reflected wave modes will lie above cutoff and therefore represent undamped modes for larger borehole diameters.

This geometric dependence produces an approximate equivalence between changes in frequency and borehole diameter (Cheng and Toksöz, 1980, 1981; Paillet, 1981). That is, an increase in borehole diameter allows an increase in the number of modes above cutoff, which is equivalent to increasing frequency to a value above cutoff for the same modes. The equivalence becomes exact in a fluid-filled borehole without logging tool when the period Equation 4.31 contains the borehole radius only in the product $m_i R$ ($i=1,2,3$).

## 4.4 BOREHOLE FORCING AND SOURCE SPECIFICATION

In calculating the properties of various modes, we assume that the source term $P_0(r,z,t)$ is zero and search for possible solutions to the period equation. In the more general situation, there will be a finite source term. We assume that the spectrum of the source can be expressed as the product of two terms:

$$P_0(r,\kappa,\omega) = S(r,\kappa)X(\omega) \qquad 4.34$$

This allows us to evaluate Fourier coefficients for given values of wavenumber and frequency in Equation 4.34. By convention, we perform the wavenumber inversion first for a normalized source function. This procedure yields Fourier coefficients for the pressure, $P_1(r,z,\omega)$, in the borehole fluid resulting from excitation by a given spatial distribution of source energy, $S(r,z)$. These coefficients may then be multiplied by various source spectra, $X(\omega)$, prior to inversion in time.

Throughout the recent synthetic microseismogram literature, the spatial representation of borehole forcing is taken as a point source. The spatial representation of this source is given by

$$S(r,\kappa) = K_0\left(m_1 r\right)e^{i\kappa - i\omega t}$$

$$u_r(r,\kappa) = \frac{\partial S}{\partial r} = -m_1 K_1\left(m_1 r\right)e^{i\kappa z}e^{-i\omega t} \qquad 4.35a$$

$$P(r,\kappa) = -\rho_f \omega^2 S = -\rho_f \omega^2 K_0\left(m_1 r\right)e^{i\kappa z}e^{-i\omega t}$$

Source representation in the case of nonaxisymmetric borehole forcing is somewhat more complicated. Winbow (1985) shows that a ring source of radius $b$ and consisting of an azimuthal distribution of intensity defined by the angular coordinate $\theta$ can be expressed as a series of responses from point sources distributed around the ring. Further analysis shows that this representation can be given as a Fourier series based on azimuthal periodicity:

$$\phi_0(r,\theta,z,t) = \sum_n \psi_n J_n(m_1 r)\cos(n\theta)e^{i(\kappa z - \omega t)} \quad b < r \leq R \qquad 4.35b$$

This solution applies in the borehole fluid in the region where $r > b$. We assume that any arbitrary nonaxisymmetric pressure source can be expressed as a similar Fourier series summing terms of successively higher azimuthal periodicity. Nonaxisymmetric forcing is modeled using source functions of the form corresponding to each individual mode in this series:

$$S(r,\kappa) = K_n(m_1 r)e^{i(\kappa z - \omega t)}$$

$$u_r(r,\kappa) = -m_1 K_n'(m_1 r)e^{i(\kappa z - \omega t)}$$

$$= \left[ -K_{n-1}(m_1 r) - \frac{n}{m_1 r} K_n(m_1 r) \right]e^{i(\kappa z - \omega t)} \qquad 4.35c$$

$$P(r,\kappa) = -\rho_f \omega^2 K_n(m_1 r)e^{i(\kappa z - \omega t)}$$

This source is consistent with the series expression for an individual azimuthal coefficient of periodicity $n$ in the series representation given by Winbow (1985). In almost all synthetic microseismogram applications we are interested in computing the borehole response to a specific azimuthal mode. In the general case of an arbitrary source, it is not very difficult to express the source as a Fourier series as in Equation 4.35b, and the general solution can be expressed as a linear combination of the microseismograms computed for each value of $n$.

Although point sources are nearly universal in the borehole model literature, various spectra have been used to model the time dependence of the pressure field produced by an acoustic logging tool. We use three of these representations to model sources in this monograph. One of the most common source spectra is given by Tsang and Rader (1979):

$$S(\omega) = \frac{8\alpha\omega_0(\alpha - i\omega)}{\left[(\alpha - i\omega)^2 + \omega_0^2\right]^2} \qquad 4.35d$$

This source represents an exponentially decaying modulation of a harmonic oscillation at the frequency $\omega_0$, where $\alpha$ and $\omega_0$ specify the width and centerband frequency of the spectrum. This particular source function was developed as a useful approximation to the response of piezoelectric and magnetostrictive borehole sources, where the source resonates at a particular frequency while the excitation energy decays in time. The parameter $\alpha$ is used to adjust the bandwidth of the spectrum.

Other sources sometimes used are the Ricker wavelet of the form (Aki and Richards, 1980)

$$S(\omega) = \left(\frac{\omega}{\omega_0}\right)^2 e^{i(\omega/\omega_0)^2} \qquad\qquad 4.35e$$

and the Kelly source (Kelly et al., 1976):

$$S(\omega) = \left(\frac{\pi}{\alpha}\right)^{1/2} \omega^3 e^{-\omega^2/4\alpha} \qquad\qquad 4.35f$$

The bandwidth of these two sources cannot be adjusted independently of the centerband frequency. However, these sources have the advantage of being "zero phase" wavelets, where the source function can be time-shifted so that $X(t) = 0$ at $t = 0$. The Ricker wavelet and Kelly source are not closely related to the known response of piezoelectric sources, but have been used extensively in the seismic literature to represent impulsive sources, and their mathematical properties are well known. In particular, the zero phase property allows the arrival of an individual frequency component of source energy at a receiver to be associated with a specific starting time. In contrast, the "sharp leading edge" of the Tsang and Rader (1979) source spectrum introduces complications in the inversion of spectra using finite difference techniques.

## 4.5    SYNTHETIC MICROSEISMOGRAM CALCULATIONS

The mode content of waveforms can be interpreted from seismic velocities and borehole geometries using the period equations given in the preceding sections. Evaluation of synthetic microseismograms produced by an acoustic logging system is carried by formulating the boundary conditions in Equations 4.24 to 4.27 using the source term, $P_0(r,z,t)$ in Equation 4.29. Incorporating the source pressure field in the boundary conditions adds a nonzero column vector to the right hand side of Equations 4.25 to 4.28. The mathematical representation of the coefficients, $\theta_{ij}$, and the elements appearing in the source column vector, $S_j$ in the solution of these nonhomogeneous matrix equations are given in Appendix 4.1.

We illustrate the general solution for the relatively simple situation where there is no logging tool in the borehole, and waves are excited by a normalized point source located along the borehole axis. The transform of the axisymmetric pressure field produced by such a point source in the borehole fluid can be represented in the form

$$P_0(r,\kappa,\omega) = X(\omega)K_0(m_1 r)e^{i\kappa z - i\omega t} \qquad 4.36a$$

The transformed pressure field, $P_1(r,\kappa,\omega)$, induced by this source is expressed in terms of the scalar potential for the displacements in the borehole fluid:

$$P_1 = -\rho_f \omega^2 \frac{\partial^2 \phi}{\partial t^2} \qquad 4.36b$$

In this situation, the coefficient $D_2$ is required to be zero because solutions must remain finite at $r = 0$. Solution of the matrix equations presented in Appendix 4.1 for the coefficient $D_1$, substitution in Equation 4.35, and Fourier inversion yield a solution of the form

$$P(r,z,t) = P_0(r,z,t) + P_1(r,z,t)$$

$$P_1(r,z,t) = \int_{-\infty}^{\infty} X(\omega)e^{-i\omega t}d\omega \int_{-\infty}^{\infty} \frac{N(\kappa,\omega)}{D(\kappa,\omega)}e^{i\kappa z}d\kappa \qquad 4.37$$

where

$$N = gK_1(m_1 r) - K_0(m_1 r) \qquad 4.37a$$

$$D = gI_1(m_1 r) + I_0(m_1 r)$$

$$g = \frac{m_1 \rho_s}{m_2 \rho_f}\left\{\left[\frac{2\kappa^2 V_s^2}{\omega^2}\right]\frac{K_0(m_2 R)}{K_1(m_2 R)} + \frac{2V_s^2 m_2 m_3}{\omega^2}\left[\frac{1}{m_3 R} + \frac{2\kappa^2 V_s^2}{\omega^2}\frac{K_0(m_3 R)}{K_1(m_3 R)}\right]\right\}$$

Appropriately configured branch cuts are required for the evaluation of $m_2$ and $m_3$ in Equation 4.36, but complex square root sign conventions in evaluating $m_1$ do not influence the calculations in these equations. The particular choice of the two possible values for the complex square root in the definition of $m_1$ does not matter because both choices are explicitly retained in the general solution. The complex square root in the definition of $m_1$ does introduce a branch line into the evaluation of the integral because some convention for taking the complex square root must be selected. Kurkjian (1985) demonstrates that the contribution introduced by the differences in value of the integrand in Equation 4.36 on either side of the fluid branch cut is exactly offset by the direct source contribution

$P_0(r,z,t)$ in Equation 4.36. This happens no matter what choice of branch cut is made. However, the choice of branch cut does matter for the other two complex square roots, where the complex root and Bessel function representations for the solution must be selected so that the radiation conditions are satisfied everywhere along the integration path. In this case, differences in values of the integrand on either side of the branch line contribute to the solution, and correct evaluation of Equation 4.36 requires that care be taken in deforming the integration path in carrying out various asymptotic representations.

Evaluation of branch cut integrals and computations of residues at singularities once were the primary means for determining solutions to the integral in Equation 4.36 (Peterson, 1974; Tsang and Rader, 1979). Modern computer software packages contain all of the formal complex variable logic to allow direct integration without resort to any other manipulations. For the rest of this chapter we will simply assume that the numerical logic in the computer respects all required branch cut conventions, yielding a numerically accurate waveform for each set of specified source conditions. The numerical integration of Equation 4.36 proceeds as in the evaluation of Equation 3.1. The integration is performed over all wavenumbers for each frequency in the source band. The Fourier coefficients determined for these frequencies are then inverted to give the synthetic waveform after convolution with the appropriate source spectra, $X(\omega)$. This approach includes two important procedures: (1) representing the double Fourier integral as a discrete sum and (2) modifying the integrand so as to avoid integrating through singularities on the real wavenumber axis. White and Zechman (1968) demonstrate that conversion of the wavenumber integral is equivalent to borehole forcing with an infinite series of sources spaced along the borehole at intervals of $\Delta z = 2\pi/\Delta\kappa$. One needs to choose $\Delta\kappa$ small enough that the waves excited by only one of these periodically spaced sources arrives in the computational time window (White and Zechman, 1968; White, 1965). The upper limit on the wavenumber integration is determined by using a convergence criterion.

The removal of the singularities from the integration path can be accomplished by adding a small amount of attenuation to the wave propagation problem. This is done by allowing for a small imaginary part to the frequency for each wavenumber integration:

$$\omega = \omega_R + i\omega_I \qquad \omega_I \geq 0 \qquad\qquad 4.37b$$

The undamped solution to Equation 4.36 can then be recovered after Fourier inversion by multiplying each value in the computed time series by the factor $\exp(\omega_I t)$. The only restriction on this method is the requirement that the imaginary part of the frequency be kept small enough so that ordinary numerical errors are not disproportionately amplified by this multiplication. The inclusion of a complex frequency also serves to offset the effects of converting the frequency integration Equation 4.36 to a finite sum. This conversion is equivalent to assuming a periodic source function. The damping introduced by the complex frequency suppresses contributions from these "ghost" sources which appear to have been triggered at earlier times (White, 1965).

An alternate approach is to directly account for the natural attenuation of real rocks and borehole fluid. In this case, frequency values used in the the inversion of Equation 4.36 are real, but complex seismic velocities are used in the computation. The use of complex velocities is equivalent to the assumption of visco-elastic properties for the solid and fluid. The real and imaginary parts of the velocities used in the computations are related to the seismic velocities and attenuation ($Q_p, Q_s, Q_f$) assumed for either seismic waves in rock or the borehole fluid according to the relation

$$\frac{1}{V_{jc}} = \frac{1}{V_j(\omega)}\left[1 + \frac{i}{2Q_j}\right]$$

4.37c

where $V_{jc}$ is the complex seismic velocity, and $V_j$ and $Q_i$ are either $V_p$, $V_s$, or $V_f$, and $Q_p, Q_s$, or $Q_f$ at the reference frequency, $\omega_0$. The use of $Q_p$ and $Q_s$ to represent seismic attenuation in Equation 4.39 is well founded, but Schoenberg et al. (1987) and Burns (1988) have questioned whether viscous borehole fluid can be characterized by use of a bulk attenuation factor $1/Q_f$ as discussed in Chapter 1. The attenuation introduced into the computation by the use of these complex velocities removes the tube wave and pseudo-Rayleigh singularities from the integration path, allowing direct numerical integration of Equation 4.34 without requiring the use of complex frequencies, or deformation from the integration path implied by Equation 4.34.

## 4.6    AN EXAMPLE OF WAVEFORM COMPUTATION

The application of synthetic borehole microseismograms to the interpretation of formation properties will be discussed in detail in the next chapters. We present a single example here to demonstrate the steps in synthetic microseismogram computation. Various intermediate calculations in waveform computation for the sandstone model used by Biot (1952) are shown in Figures 4.3 and 4.4. The computations illustrate that there are three undamped modes present: the tube wave and the two pseudo-Rayleigh modes with cutoff frequencies within the source frequency band. The borehole response function $D_1$ $I_0(m_1 r)$ in Equation 4.13 in the frequency-wavenumber plane is shown in Figure 4.3. Note that this representation includes multiplication by the factor $\rho_f \omega^2$ introduced by the second derivative of the potential in Equation 4.10. Some of the other references in the literature do not include this factor, indicating a very different form for the response function for the same solution. The response function illustrates the Stoneley and pseudo-Rayleigh modes, and the discontinuity associated with critically refracted compressional head waves along the lines $\omega/\kappa = V_p$ and $\omega/\kappa = V_s$, which represent the trajectory of the branch points in the $\omega$-$\kappa$ plane. On the scale of these plots, the variations in amplitude of the response function near these lines is obscured by the merging of the pseudo-Rayleigh modes with the branch cut lines in $\omega$–$\kappa$ space. The expanded plots would show the same changes in response function amplitude with small changes of frequency illustrated in Figure 3.7.

The first part of Figure 4.4 illustrates the relative excitation of the three modes (tube wave and two pseudo-Rayleigh modes) determined from the wavenumber integration of Equation 4.35 for discrete wavenumbers. The second part of the Figure illustrates the source spectrum convolved with the calculated spectra for these modes. Comparison of the source spectrum with the mode excitation functions indicates that the source spectrum is exciting the Airy phase of the lowest pseudo-Rayleigh mode. The composite waveform calculated by evaluating Equation 4.34 for this case (Figure 4.5) at two different receiver offsets indicates that the maximum energy in the lowest pseudo-Rayleigh mode arrives as a single pulse of energy after direct fluid travel time, as expected from the group velocity calculations.

## 4.7 MODE ATTENUATION AND PARTITION COEFFICIENTS

The early success of acoustic logging in characterizing the vertical distribution of compressional velocity encouraged interest in using acoustic logs to characterize other acoustic properties of rocks. Attenuation of compressional waves ($Q_p$) appeared to be another seismic property of interest to the geophysicist which could be estimated on the basis of the decrease in amplitude of measured acoustic signals with source-receiver separation in the borehole. However, the mathematical theory of head wave propagation along the borehole indicates that the attenuation of critically refracted head waves depends upon outward radiative losses, in addition to intrinsic attenuation. For example, a broad-band acoustic source will excite frequencies spanning more than one of the cut-off frequencies for the critically refracted compressional modes, resulting in the superposition of compressional wave excitation peaks with different rates of radiation loss as the waves propagate along the borehole. The same arguments apply in the case of shear head wave propagation. Initial attempts to relate measured decreases in head wave amplitude at different source-receiver spacings were not very successful (Willis, 1983). Calculation of radiation losses for head waves seems to be very sensitive to the properties of the source spectra in the vicinity of mode cutoffs so as to greatly complicate the delicate process of accounting for all energy losses when interpreting waveform logs. Measured changes in head wave amplitude at different source-receiver spacing are at best a qualitative means of characterizing intrinsic attenuation in rocks surrounding the borehole.

The guided pseudo-Rayleigh and tube wave modes are not subject to attenuation by outward radiation during propagation. Such modes seem ideally

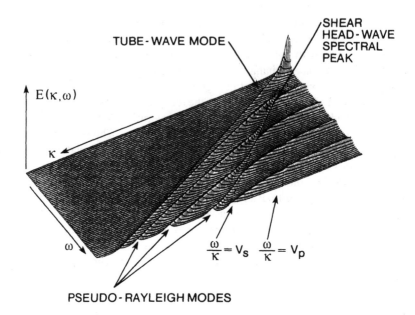

FIGURE 4.3.  Pressure response function for the fluid-filled borehole plotted as a third dimension in frequency and wavenumber space, illustrating spectral peaks associated with tube-wave and pseudo-Rayleigh modes, and location of regions where shear head waves are excited with maximum amplitude; lithology is sandstone; fluid is water, with $R = 12$ cm, $V_p = 4.8$ km/s, $V_s = 2.8$ km/s, and $\rho_s = 2.4$ g/cm$^3$.

suited for the measurement of attenuation because the decrease in mode amplitude can be directly related to intrinsic attenuation as long as the dispersive properties of the modes are accounted for. In practice, this compensation amounts to careful filtering in frequency space and adjustment of time windows to allow for the dispersion. The interpretation of mode attenuation then reduces to relating the observed decrease in amplitude to the intrinsic attenuation of the rock ($Q_p$, $Q_s$) and the borehole fluid ($Q_f$).

Mode attenuation can be related to intrinsic attenuation by expanding mode attenuation ($Q_m$) in terms of the partial derivatives of the mode velocities ($V_m$) (Anderson and Archambeau, 1964):

$$\frac{1}{Q_m} = \frac{V_p}{V_m}\frac{\partial V_m}{\partial V_p}\frac{1}{Q_p} + \frac{V_s}{V_m}\frac{\partial V_m}{\partial V_s}\frac{1}{Q_s} + \frac{V_f}{V_m}\frac{\partial V_m}{\partial V_f}\frac{1}{Q_f} \qquad 4.38$$

This expression neglects second and higher terms in ($1/Q$). Cheng et al. (1982) showed that this error is less than 10% for values of $Q_m$ greater than 10.0. The terms multiplying $Q_p^{-1}$, etc., in this expression are denoted as partition coefficients because they describe the way in which the observed attenuation of the mode is represented as a linear combination of the attenuation in the rock and fluid.

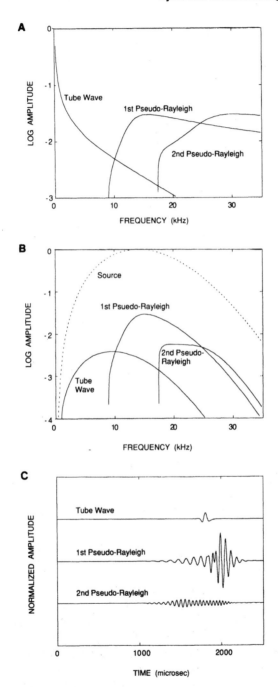

FIGURE 4.4.   Pseudo-Rayleigh and tube-wave modes in a typical sandstone: (A) mode excitation compared to source spectrum, (B) mode excitation convolved with source, and (C) synthetic microseismograms for individual modes excited by the source; fluid is water, $R = 10.2$ cm, $V_p = 4.57$ km/s, $V_s = 2.74$ km/s, and $\rho_s = 2.3$ g/cm$^3$; $f_0 = 15$ kHz.

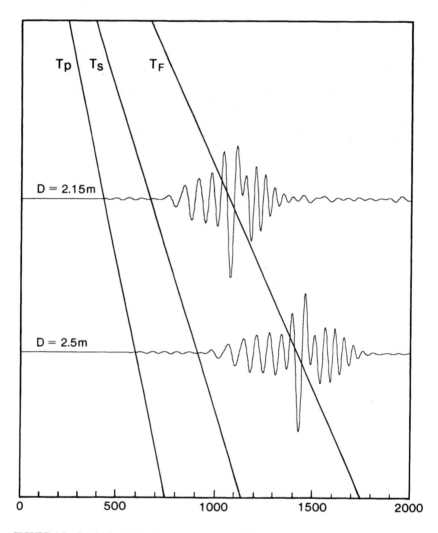

FIGURE 4.5. Synthetic microseismograms at two different receiver offsets computed for the sandstone illustrated in Figure 4.4; fluid is water, $R = 10.2$ cm, $V_p = 4.57$ km/s, $V_s = 2.74$ km/s, and $\rho_s = 2.3$ g/cm$^3$; $f_0 = 15$ kHz.

Cheng et al. (1982) computed the partition coefficients for tube wave and pseudo-Rayleigh modes using the elastic constants for rock and the borehole fluid, and the wave equation solutions for displacements. The calculation is based on the representation of the strain energy density of the modes, and the fact that the Langrangian (difference of average kinetic energy and strain energy densities) is stationary for these modes (Aki and Richards, 1980). If the frequency, elastic constants, and density are perturbed and wavenumber held constant, Cheng et al. (1982) give

$$\frac{1}{Q} = \frac{1}{2\omega^2 I}\left\{\int_0^\infty (\lambda + 2\mu)\left(\frac{du_1}{dr} + \frac{u_1}{r} - \kappa u_2\right)^2 \frac{1}{Q_p} r dr \right.$$

$$+ \int_R^\infty 2\mu\left[\left(\frac{du_1}{dr}\right)^2 + \left(\frac{u_1}{r}\right)^2 + \kappa^2 u_2{}^2 - \left(\frac{du_1}{dr} + \frac{u_1}{r} - \kappa u_2\right)^2\right.$$

$$\left.\left. + \frac{1}{2}\left(\kappa u_1 + \frac{du_2}{dr}\right)^2\right]\frac{1}{Q_s} r dr\right\}$$ 

4.39

$$I = \frac{1}{2}\int_0^\infty \rho\left(u_1{}^2 + u_2{}^2 r dr\right)$$

with $u_1$ and $u_2$ given by

$$u_r = u_1(r)e^{i(\kappa z - \omega t)}$$

$$u_z = iu_2(r)e^{i(\kappa z - \omega t)}$$

where $u_r$ and $u_z$ are the radial and axial components of displacement in the fluid or solid. Cheng et al. (1982) separated the integral in Equation 4.4 into segments over the fluid and rock, and then substituted the expressions for displacements $u_r$ and $u_z$ given in Section 4.3 into these equations. The results allow calculation of partition coefficients as a function of frequency for the tube wave and pseudo-Rayleigh modes for any given set of elastic constants and intrinsic attenuation values ($1/Q$) for fluid and rock. Examples of partition coefficients for typical rocks are given in Figure 4.6. The results for limestone and granite are similar, even though the seismic velocities are somewhat different. The shear attenuation ($Q_s$) of the rock almost completely determines the pseudo-Rayleigh mode attenuation in the frequency range just above mode cutoff, while the attenuation of the borehole fluid dominates mode attenuation at frequencies greater than the frequency corresponding to the minimum in group velocity (Airy phase). In contrast, the tube wave mode attenuation is dominated at all frequencies by the fluid attenuation ($Q_f$). However, there is a significant difference in the partition coefficients between granite and limestone; the shear attenuation has an important contribution to mode attenuation at lower frequencies in the limestone.

In the case of rocks with lower velocities, the velocity contrast between the borehole fluid and the surrounding rock is reduced, and there is a broad transition

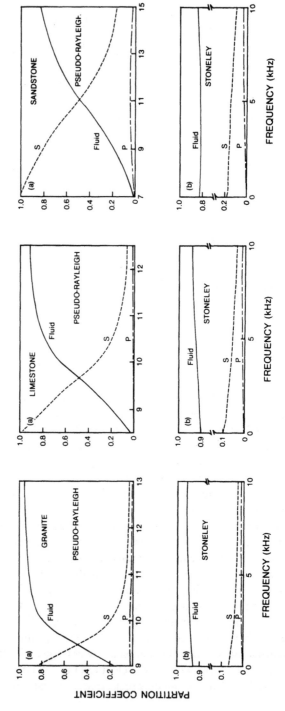

FIGURE 4.6.   Compressional, shear, and fluid partition coefficients calculated for attenuation in (A) granite, (B) limestone, and (C) sandstone.

between lower mode frequencies, where attenuation is dominated by shear attenuation, and higher frequencies, where attenuation is dominated by fluid attenuation. For example, the partition coefficients for sandstone indicate that there is a significant contribution from shear attenuation to total mode attenuation even at frequencies greater than the frequency at the minimum in mode group velocity. At the same time, shear attenuation provides a significant contribution to tube wave attenuation at all frequencies. The shear contribution to tube wave attenuation becomes substantial for softer sediments or unconsolidated sandstone in which shear velocity falls below the acoustic velocity in the borehole fluid. In that case, radiation conditions do not permit pseudo-Rayleigh modes. The partition coefficients for a typical soft sediment indicate that shear attenuation dominates the tube wave attenuation at all but the lowest frequencies.

## 4.8    SYNTHETIC MICROSEISMOGRAMS FOR NONSYMMETRIC MODES

Although most early interest in synthetic microseismograms and their applications in the interpretation of acoustic waveform logs was directed towards azimuthally symmetric sources, recent interest in shear wave logging has focused on the interpretation of nonsymmetric modes. One of the earliest attempts to compute synthetic microseismograms for nonsymmetric modes was by White (1967). The first application of nonsymmetric sources in shear logging systems were based on the assumption that such sources would primarily excite shear body waves in the formation, so that compressional arrivals would be of very small amplitude and would not interfere with determination of shear velocities (Kitsunezaki, 1980; Williams et al., 1984; Chen, 1988, 1989; Kaneko et al., 1989). However, mode generation mechanisms similar to those encountered for symmetric modes need to be considered in the interpretation of waveforms excited by nonsymmetric sources. The nonsymmetric modes are found to have phase and group velocities near $V_s$ because the departure from azimuthal symmetry fundamentally changes the relation between mode propagation and the constructive interference of waves in the borehole fluid. In the limit of low frequencies (approaching mode cutoff), nonsymmetric modes are found to approach simple shear body waves in character (Winbow, 1985; Kurkjian and Chang, 1986). Even in that limit, the identity of recorded wave arrivals as trapped modes and not simple shear body waves cannot be ignored in the analysis.

The calculation of synthetic borehole microseismograms can be expanded to include the effect of nonsymmetric sources located on the borehole axis, or for symmetric sources located off of the borehole axis by modifying the separation of variable solution to include azimuthal dependence. In order to do this, we have to allow for horizontally polarized shear waves in addition to the vertically polarized shear waves represented by the shear potential in the axisymmetric case. The additional scalar potential is introduced using the representation (Kurkjian and Chang, 1986)

$$\boldsymbol{u} = \nabla\phi + \nabla\times\zeta_1{}^{SH} + \nabla\times\nabla\times\zeta_2{}^{SV} \qquad\qquad 4.40$$

where both $\zeta_1$ and $\zeta_2$ consist of single components $\psi$ and $\Gamma$, and are associated with horizontal (*SH*) and vertical (*SV*) motion, respectively, in the $z$ direction. With this representation, the expression for the displacements in the solid become

$$
\begin{aligned}
u_r &= \frac{\partial\phi}{\partial r} + \frac{1}{r}\frac{\partial\psi}{\partial\theta} + \frac{\partial^2\Gamma}{\partial r\partial z} \\
u_\theta &= \frac{1}{r}\frac{\partial\phi}{\partial r} - \frac{\partial\psi}{\partial r} + \frac{1}{r}\frac{\partial^2\Gamma}{\partial\theta\partial z} \\
u_z &= \frac{\partial\phi}{\partial z} - \frac{\partial^2\Gamma}{\partial r^2} - \frac{1}{r}\frac{\partial\Gamma}{\partial r} - \frac{1}{r^2}\frac{\partial^2\Gamma}{\partial\theta^2}
\end{aligned}
\qquad 4.41
$$

Restricting synthetic microseismograms to the case of a fluid-filled borehole without logging tool, the solutions in the rock and borehole fluid take the form

$$
\begin{aligned}
\phi &= \left[A_1 K_n\!\left(m_2 r\right) + A_2 I_n\!\left(m_2 r\right)\right]\cos(n\theta)e^{i(\kappa z - \omega t)} \\
\psi &= \left[E_1 K_n\!\left(m_3 r\right) + E_2 I_n\!\left(m_3 r\right)\right]\cos(n\theta)e^{i(\kappa z - \omega t)} \\
\Gamma &= \left[B_1 K_n\!\left(m_3 r\right) + B_2 I_n\!\left(m_3 r\right)\right]\sin(n\theta)e^{i(\kappa z - \omega t)}
\end{aligned}
\qquad 4.42
$$

Previous authors have used a reference angle, $\theta_0$, to define the azimuthal argument in these expressions. We have generalized that expression by defining $\theta$ with respect to that reference angle. This does not restrict the generality of the solution, but constrains the angular coordinate that is used to match solutions with a given multipole source.

The potential for displacements in the borehole fluid remains the same as in Equation 4.12 except for an additional factor accounting for azimuthal periodicity:

$$\phi = \left[D_1 I_n\!\left(m_1 r\right) + D_2 K_n\!\left(m_1 r\right)\right]\cos(n\theta)e^{i(\kappa z - \omega t)} \qquad 4.43$$

As in previous computations for the fluid-filled borehole, we require the coefficients $A_2, B_2, D_2,$ and $E_2$ to vanish in order that the solution be defined along the borehole axis and at large radii. The four remaining unknown Fourier coefficients in these expressions are determined by invoking the boundary conditions at the borehole wall. These are continuity of radial displacement and radial stress, and vanishing azimuthal and axial components of the shear stress at the borehole wall. This yields a set of four coupled equations analogous to Equation 4.32. We have gone from two to four coupled equations for the fluid-filled borehole without logging tool by addition of the horizontal shear mode, and because we were able to reduce the system in Equation 4.29 by elimination of one variable (i.e., using the vanishing of azimuthal shear stress at the borehole wall).

Substituting the solution of Equation 4.42 into the expressions for the displacements and stresses in the boundary conditions yields a set of coupled equations of the form

$$
\begin{pmatrix}
\theta_{11} & \theta_{12} & \theta_{13} & 0 \\
\theta_{21} & \theta_{22} & \theta_{23} & 0 \\
0 & \theta_{32} & \theta_{33} & \theta_{34} \\
0 & \theta_{42} & \theta_{43} & \theta_{44}
\end{pmatrix}
\begin{pmatrix}
D_1 \\
A_1 \\
B_1 \\
E_1
\end{pmatrix}
=
\begin{pmatrix}
S_1 \\
S_2 \\
0 \\
0
\end{pmatrix}
\qquad 4.44
$$

where the expressions for the coefficients, $\theta_{ij}$, and the source vector are listed in Appendix 4.2. Synthetic microseismograms are computed by evaluation of the integral in Equation 4.37, except that the integrand is determined as function of frequency and wavenumber by solving the system of Equation 4.44 rather than Equation 4.29. The expression for the nonsymmetric pressure source used in the appendix is based on the mathematical expressions for the far-field form of the nonsymmetric pressure field produced by one or more pairs of point sources (monopoles) symmetrically arrayed around the borehole axis (Kurkjian and Chang, 1986; Schmitt et al., 1988b). However, one can expand any assumed nonsymmetric source in terms of a cosine series and impose such an assumed pressure field on the borehole wall to compute borehole excitement with any nonsymmetric source.

The sometimes confusing reference to logging with flexural modes in the literature indicates that some clarification of terminology is required. In this study, the terms "dipole" and "quadrupole" refer to the number of point sources used to generate the azimuthal distribution of energy provided by the logging source (Kurkjian and Chang, 1986; Winbow, 1988; Baker and Winbow, 1988). In that sense, the order of the "pole" in the source will be twice the value of $n$ used in Equation 4.42. The term "flexural" mode has been applied to source transducers producing a sideways forcing of the displacement at the surface of the logging tool as proposed by White (1967). This type of transducer is intended to excite the $n = 1$, $m = 0$ (dipole) mode, but probably introduces small amounts of energy into other, higher order azimuthal and radial modes.

The period equation for the nonsymmetric modes (Det $(\theta_{ij}) = 0$ in Equation 4.44) yields the characteristic velocities and cutoff frequencies for the modes which exist for various values of $n$, the azimuthal periodicity, and $m$, the order of the individual poles within the infinite series of poles for each $n$. Examples of the lowest ($m = 0$) modes for the $n = 1$ (dipole) and $n = 2$ (quadrupole) cases are

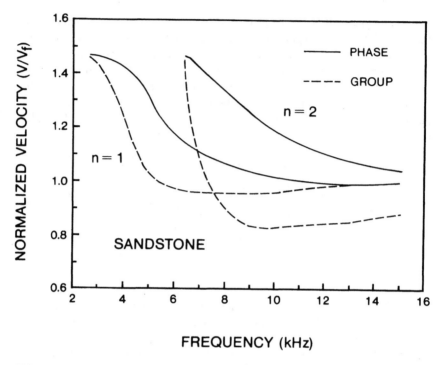

FIGURE 4.7. Phase and group velocities for the fundamental ($m=0$) mode of the nonsymmetric dipole ($n=1$) and quadrupole ($n=2$) modes; sandstone lithology; fluid is water, $R = 10$ cm, $V_p = 4.0$ km/s, and $\rho_s = 2.3$ g/cm$^3$.

illustrated for a sandstone lithology in Figure 4.7. The nonsymmetric wave solutions allow modes analogous to the tube wave for $n > 1$ and $m = 0$, and analogous to the pseudo-Rayleigh modes when $n > 1$ and $m > 1$. The $n > 1$, $m = 0$ modes are similar to the tube wave ($n = 0, m = 0$) in that there is no low frequency cutoff, and the phase and group velocities approach the Stoneley velocity in the high frequency, short wavelength limit. However, the $n = 1$ and $n = 2$ modes in Figure 4.7 appear to have a low frequency cutoff for $m = 0$ because mode excitation falls off rapidly at frequencies less than 3 kHz. Similar dispersion and mode excitation curves are given for the $n = 1$ and 2, $m = 0$ modes in Figure 4.8 for a "slow" shale formation where nonsymmetric waveform logging is especially useful in measuring $V_s$.

The appearance of nonsymmetric waveforms at frequencies below the cutoff for the $m = 1$ mode are illustrated in Figures 4.9 (sandstone) and 4.10 (shale). These figures present the $n = 1$ (dipole) and $n = 2$ (quadruple) modes. Synthetic microseismograms computed for a source frequency range which is entirely below cutoff frequency for the $m = 1$ and higher modes is completely dominated by the $m = 0$ mode, just as the tube wave dominates at low

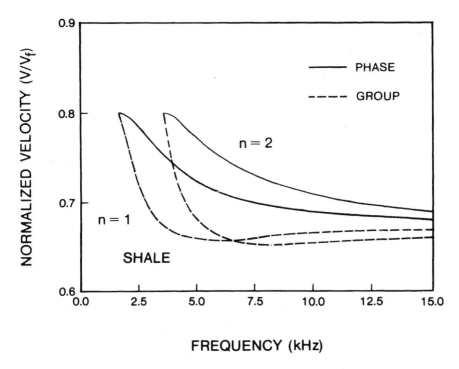

FIGURE 4.8. Phase and group velocities for the fundamental ($m=0$) mode of the nonsymmetric dipole ($n=1$) and quadrupole ($n=2$) modes; shale lithology; fluid is water, $R = 10$ cm, $V_p = 2.2$ km/s, $V_s = 1.2$ km/s, and $\rho_s = 2.1$ g/cm$^3$.

frequencies for the azimuthally symmetric case. At frequencies slightly below cutoff for the $m = 1$ mode, head waves begin to appear in the computed microseismograms, and the group velocity of the $m = 0$ mode appears to fall below $V_s$. Mode content in the $n = 1$ case is illustrated in a series of waveforms computed over the frequency range from 0.5 to 10.5 kHz (Figure 4.11). The figure shows the increase in relative amplitude of the $m = 1$ mode with respect to the $m = 0$ mode as frequency is increased above the cutoff for the $m = 1$ mode. These results indicate that shear wave logging with nonsymmetric sources needs to be conducted using source frequencies less than 1 kHz or much less than those corresponding to the group velocity minimum in Figure 4.7. The waveforms in Figure 4.9 indicate that the $n = 1$, $m = 0$ mode does produce arrivals traveling at formation shear velocity when source frequency is low enough. This sensitivity of shear logging systems to source frequency was not expected when the method was originally proposed (White, 1967; Kitsunezaki, 1980; Zemanek et al., 1984).

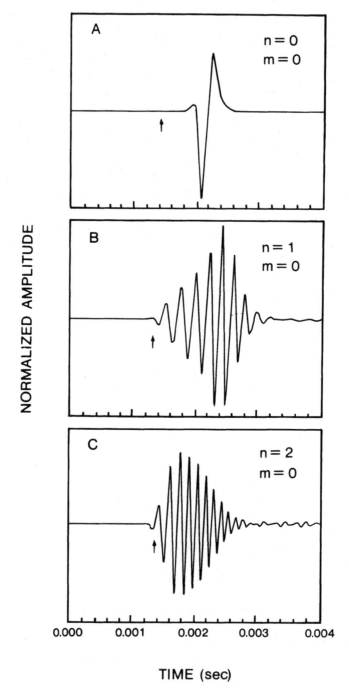

FIGURE 4.9.  Comparison of synthetic microseismograms of (A) the fundamental ($m=0$) mode for the tube-wave ($n=0$) with (B) the dipole ($n=1$) and (C) quadrupole ($n=2$) modes sandstone lithology (arrow denotes shear arrival time for critically refracted waves); shale lithology of Figure 4.7 with $f_0 = 3$ kHz.

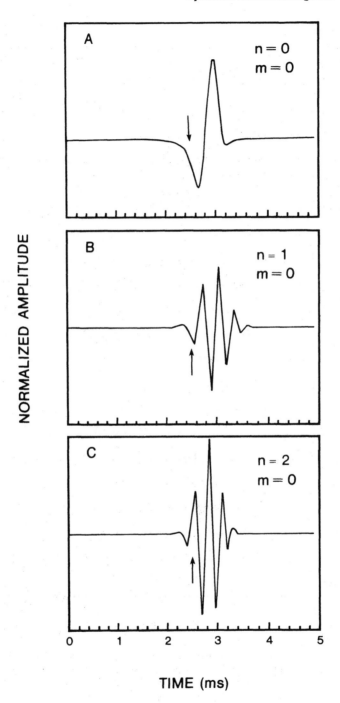

FIGURE 4.10. Comparison of synthetic microseismograms of (A) the fundamental ($m$=0) mode for the tube-wave ($n$=0) with (B) the dipole ($n$=1) and (C) quadrupole ($n$=2) modes; (arrow denotes arrival time for critically refracted shear waves); shale lithology of Figure 4.8 with $f_0$ = 3 kHz.

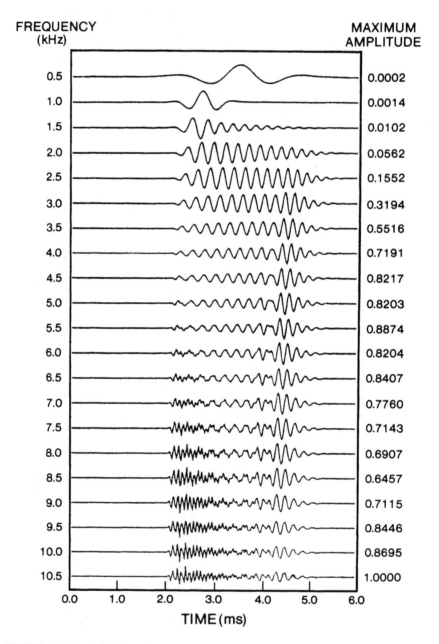

FIGURE 4.11. Synthetic microseismograms computed at frequencies ranging from 0.5 to 10.5 kHz for the dipole ($n=1$) source in sandstone, illustrating increasing mode content with increasing frequency; sandstone lithology of Figure 4.7.

## 4.9 SYMMETRIC AND NONSYMMETRIC MODES IN TRANSVERSALLY ISOTROPIC SOLIDS

Many if not most real rocks are not perfectly isotropic. However, expansion of the synthetic microseismogram calculations to include nonisotropic rocks greatly complicates the mathematics. One of the simplest cases of nonisotropic rocks that can be considered is the transversely isotropic solid. In that case we assume a bedded or layered structure in which the axis of symmetry is nearly vertical, and parallel to the borehole.

In the case of transversely isotropic formations, the generalized nonsymmetric solution for the Fourier coefficients given in Equation 4.42 has to be modified in two ways: (1) the stress-strain relations are changed to include the nonisotropic relations in Equation 1.9, and (2) the waves are expressed in terms of the quasi-$P$ and quasi-$SV$ modes in Equation 1.10 rather than directly related to individual scalar potentials. In the latter case we assume solutions of the form

$$\phi^P = \left[ A_1 K_n\left(m_{qp} r\right) + A_2 I_n\left(m_{qp} r\right) \right] \cos(n\theta) e^{i(\kappa z - \omega t)}$$

$$\Gamma^{SV} = \left[ B_1 K_n\left(m_{qs} r\right) + B_2 I_n\left(m_{qs} r\right) \right] \sin(n\theta) e^{i(\kappa z - \omega t)}$$

$$\psi^{SH} = \left[ E_1 K_n\left(m_{sh} r\right) + E_2 I_n\left(m_{sh} r\right) \right] \cos(n\theta) e^{i(\kappa z - \omega t)}$$

$$\phi_n = \phi^P + a\Gamma^{SV}$$

$$\Gamma_n = b\phi^P + \Gamma^{SV}$$

$$\psi_n = \psi^{SH} \qquad\qquad 4.45$$

and

$$u_r = \frac{\partial \phi}{\partial r} + \frac{1}{r}\frac{\partial \psi}{\partial \theta} + \frac{\partial^2 \Gamma}{\partial r \partial z}, \text{ etc.}$$

where the wavenumbers $m_{qp}$, $m_{qs}$, and $m_{sh}$ are derived from the expressions in Equation 1.10a using the relations:

$$m_{qp}^{2} = \kappa^2 - \frac{\omega^2}{V_{qp}^{2}} \qquad V_{qp} = \frac{\omega}{M}$$

$$m_{qs}^{2} = \kappa^2 - \frac{\omega^2}{V_{qp}^{2}} \qquad V_{qs} = \frac{\omega}{K}$$

and                                                                                            4.46

$$m_{sh}^{2} = \frac{L\kappa^2 - \omega^2 \rho_s}{N}$$

where $M$, $K$, $L$, and $N$ are defined in Equations 1.9 and 1.10a, and where the expressions $a$ and $b$ in Equation 1.10 give the coupling between shear and compression in the quasi-$P$ and quasi-$SV$ modes. The rest of the solution proceeds as above, using the same general expressions for the source pressure function. The algebraic manipulation required to carry this out is evidently more involved than in any of the previous examples. The matrix coefficients analogous to those in Equation 4.44 for the transversely isotropic problems are listed in Appendix 4.3.

# Appendix 4.1

## Boundary Condition Matrix Elements for Borehole Forcing with Axisymmetric Pressure Sources

The continuity of displacement and normal stresses at the borehole wall and at the surface of the logging tool give a set of four equations of the form

CONTINUITY OF RADIAL DISPLACEMENT $r=R$

CONTINUITY OF RADIAL STRESS $r=R$

CONTINUITY OF RADIAL DISPLACEMTN $r=R$

CONTINUITY OF RADIAL STRESS $r=R$

$$\begin{pmatrix} \theta_{11} & \theta_{12} & \theta_{13} & 0 \\ \theta_{21} & \theta_{22} & \theta_{23} & 0 \\ 0 & \theta_{32} & \theta_{33} & \theta_{34} \\ 0 & \theta_{42} & \theta_{43} & \theta_{44} \end{pmatrix} \begin{pmatrix} A_1 \\ D_1 \\ D_2 \\ A_4 \end{pmatrix} = \begin{pmatrix} S_1 \\ S_2 \\ S_3 \\ S_4 \end{pmatrix}$$

$$A4.1$$

where $S_1$, $S_2$, $S_3$, and $S_4$ are source terms derived from the expression for $u_r$ and $P_0$ at the borehole wall ($r=R$) and the surface of the logging tool ($r=R_0$):

$$S_1 = -m_1 K_1 (m_1 R)$$

$$S_2 = -\rho_f K_0 (m_1 R)$$

$$S_3 = -m_1 K_1 (m_1 R_0)$$

$$S_4 = -\rho_f K_0 (m_1 R_0)$$

$$A4.2$$

The values of the nonzero elements in the matrix of coefficients, $\theta_{ij}$, are (with the common factor $\exp[i(\kappa x - \omega t)]$ eliminated):

$$\theta_{11} = \frac{\omega^2}{2\kappa^2 V_s^2 - \omega^2} m_2 K_1 (m_2 R)$$

$$\theta_{12} = -m_1 I_1 (m_1 R)$$

$$\theta_{13} = m_1 K_1 (m_1 R)$$

$$\theta_{21} = \rho_s \left\{ \frac{2\kappa^2 V_s^2 - \omega^2}{\omega^2} K_0 (m_2 R) + \frac{2 V_s^2 m_2 m_3 K_1 (m_2 R)}{\omega^2 - 2\kappa^2 V_s^2} \left[ \frac{1}{m_3 R} + \frac{2\kappa^2 V_s^2 K_0 (m_3 R)}{\omega^2} \frac{K_0(m_3 R)}{K_1 (m_3 R)} \right] \right\}$$

$$\theta_{22} = \rho_f I_0 (m_1 R)$$

$$\theta_{32} = m_1 I_1\left(m_1 R_0\right)$$

$$\theta_{33} = m_1 K_1\left(m_1 R_0\right)$$

$$\theta_{34} = \frac{\omega^2}{2\kappa^2 V_{st}^{\ 2} - \omega^2} m_4 K_1\left(m_4 R_0\right)$$

$$\theta_{42} = \rho_f I_0\left(m_1 R_0\right)$$

$$\theta_{43} = \rho_f K_0\left(m_1 R_0\right)$$

$$\theta_{44} = \rho_t \left\{ \frac{2\kappa^2 V_{st}^{\ 2} - \omega^2}{\omega^2} I_0\left(m_4 R_0\right) + \frac{2 V_{st}^{\ 2} m_4 m_5 I_1\left(m_4 R_0\right)}{\omega^2 - 2\kappa^2 V_{st}^{\ 2}} \left[ \frac{1}{m_5 R_0} + \frac{2\kappa^2 V_{st}^{\ 2} I_0\left(m_5 R_0\right)}{\omega^2 \quad I_1\left(m_5 R_0\right)} \right] \right\}$$

where subscripts $f$, $s$, $t$ denote properties of fluid, solid, and steel tool; subscripts 1, 2, 3, 4, 5 on radial wavenumbers correspond to $V_f$, $V_p$, $V_s$, $V_{pt}$, and $V_{st}$ (the compression and shear velocities of steel).

In situations where the logging tool is assumed rigid, the boundary condition equations are reduced to the set

$$\begin{pmatrix} \theta_{11} & \theta_{12} & \theta_{13} \\ \theta_{21} & \theta_{22} & \theta_{23} \\ 0 & \theta_{32} & \theta_{33} \end{pmatrix} \begin{pmatrix} A_1 \\ D_1 \\ D_2 \end{pmatrix} = \begin{pmatrix} S_1 \\ S_2 \\ 0 \end{pmatrix} \qquad A4.4$$

In the most common situation, solutions are obtained for the fluid-filled borehole without logging tool, reducing the boundary condition equations to the set

$$\begin{pmatrix} \theta_{11} & \theta_{12} \\ \theta_{21} & \theta_{22} \end{pmatrix} \begin{pmatrix} A_1 \\ D_1 \end{pmatrix} = \begin{pmatrix} S_1 \\ S_2 \end{pmatrix} \qquad A4.5$$

# Appendix 4.2

## Boundary Condition Matrix Elements for Borehole Forcing with Nonaxisymmetric Pressure Sources

The continuity of radial displacement and stress at the borehole wall result in a set of four coupled equations for the amplitude of the scalar potential in the borehole fluid $(D_n)$ and the amplitudes of the scalar $(A_n)$ and two vector potentials $(B_n$ and $E_n)$ for each value of azimuthal periodicity. Note that we have dropped the numerical subscript for each of these amplitudes under the assumption that only the Bessel functions $I_n(x)$ are retained in the borehole fluid and only $K_n(x)$ are used in the surrounding formation. The boundary condition equations evaluated at the borehole wall $(r = R)$ assume the form

$$
\begin{array}{l}
\text{CONTINUITY OF RADIAL DISPLACEMENT} \\
\text{CONTINUITY OF RADIAL STRESS} \\
\text{CONTINUITY OF AZIMUTHAL SHEAR STRESS} \\
\text{CONTINUITY OF AXIAL SHEAR STRESS}
\end{array}
\begin{pmatrix}
\theta_{11} & \theta_{12} & \theta_{13} & \theta_{14} \\
\theta_{21} & \theta_{22} & \theta_{23} & \theta_{24} \\
0 & \theta_{32} & \theta_{33} & \theta_{34} \\
0 & \theta_{42} & \theta_{43} & \theta_{44}
\end{pmatrix}
\begin{pmatrix} D_n \\ A_n \\ B_n \\ E_n \end{pmatrix}
=
\begin{pmatrix} S_1 \\ S_2 \\ 0 \\ 0 \end{pmatrix}
$$

$$A4.6$$

The source terms are

$$
S_1 = m_1 K_n'(m_1 R) = -m_1 \left\{ \frac{nK_n(m_1 R)}{m_1 R} + K_{n+1}(m_1 R) \right\}
$$

$$A4.7$$

$$
S_2 = -\rho_f K_n(m_1 R) \qquad n \geq 1
$$

Using the following notation for combinations of Bessel functions:

$$
X_1(mR) = -\left[ \frac{n}{R} K_n(mR) - mK_{n+1}(mR) \right]
$$

$$
X_2(mR) = X_1(mR) - \frac{K_n(mR)}{R}
$$

$$A4.8$$

$$
X_3(mR) = X_1(mR) + \frac{n^2}{R} K_n(mR)
$$

the nonzero elements in the boundary condition matrix become

$$\theta_{11} = \frac{n}{R} I_n(m_1 R) + m_1 I_{n+1}(m_1 R)$$

$$\theta_{12} = X_1(m_2 R)$$

$$\theta_{13} = \frac{n}{R} K_n(m_3 R)$$

$$\theta_{14} = i\kappa X_1(m_3 R)$$

$$\theta_{21} = \rho_f \omega^2 I_n(m_1 R)$$

$$\theta_{22} = \rho_s V_s^2 \left\{ (\kappa^2 + m_3^2) K_n(m_2 R) - \frac{2}{R} X_3(m_2 R) \right\}$$

$$\theta_{23} = \frac{2n\rho_s V_s^2}{R} X_2(m_3 R)$$

$$\theta_{24} = 2i\kappa\rho_s V_s^2 \left\{ m_3^2 K_n(m_3 R) + \frac{1}{R} X_3(m_3 R) \right\}$$

$$\theta_{32} = -\frac{2n\rho_s V_s^2}{R} X_2(m_2 R)$$

$$\theta_{33} = -\rho_s V_s^2 \left\{ -m_3^2 K_n(m_3 R) - \frac{2}{R} X_3(m_3 R) \right\}$$

$$\theta_{34} = -\frac{2i\kappa n\rho_s V_s^2}{R} X_2(m_3 R)$$

$$\theta_{42} = 2i\kappa\rho_s V_s^2 X_1(m_2 R)$$

$$\theta_{43} = \frac{-i\kappa\rho_s V_s^2}{R} K_n(m_3 R)$$

$$\theta_{44} = -\rho_s V_s^2 (\kappa^2 + m_3^2) X_1(m_3 R)$$

A4.9

# Appendix 4.3

## Boundary Condition Matrix Elements for Borehole Forcing with Axisymmetric and Nonaxisymmetric Pressure Sources in Transversely Isotropic Formations

In order to describe the wave propagation in transversely isotropic elastic formations, the stress-strain relations are expanded to include five rather than two Lamé constants ($A, C, N, F$, and $L$) as defined in Equation 1.9. Using this notation, we calculate the boundary conditions at the borehole wall ($r=R$) using the continuity of radial displacement and radial and shear stress to solve for the amplitude of the scalar potential in the borehole fluid ($D_n$) and the amplitude of the scalar potential ($A_n$) and two vector potentials ($B_n$ and $E_n$) in the formation

CONTINUITY OF RADIAL DISPLACEMENT

CONTINUITY OF RADIAL STRESS

CONTINUITY OF AZIMUTHAL SHEAR STRESS

CONTINUITY OF AXIAL SHEAR STRESS

$$\begin{pmatrix} \theta_{11} & \theta_{12} & \theta_{13} & \theta_{14} \\ \theta_{21} & \theta_{22} & \theta_{23} & \theta_{24} \\ 0 & \theta_{32} & \theta_{33} & \theta_{34} \\ 0 & \theta_{42} & \theta_{43} & \theta_{44} \end{pmatrix} \begin{pmatrix} D_n \\ A_n \\ B_n \\ E_n \end{pmatrix} = \begin{pmatrix} S_1 \\ S_2 \\ 0 \\ 0 \end{pmatrix}$$

$$A4.10A$$

For the axisymmetric case ($n=0$), the azimuthal shear stress condition is dropped, and the boundary condition matrix becomes

$$\begin{pmatrix} \theta_{11} & \theta_{12} & \theta_{14} \\ \theta_{21} & \theta_{22} & \theta_{14} \\ \theta_{42} & \theta_{43} & \theta_{44} \end{pmatrix} \begin{pmatrix} D_n \\ A_n \\ E_n \end{pmatrix} = \begin{pmatrix} S_1 \\ S_2 \\ 0 \end{pmatrix}$$

$$A4.10B$$

The two source terms are then given by

$$S_1 = -m_1 \left\{ \frac{nK_n(m_1R)}{m_1R} + K_{n+1}(m_1R) \right\}$$

$$A4.11A$$

$$S_2 = -\rho_f K_n(m_1R) \qquad n \geq 1$$

$$S_1 = -m_1 K_1(m_1R)$$

$$S_2 = -\rho_f K_0(m_1R) \qquad n = 0 \qquad A4.11B$$

The values of the nonzero elements in the boundary condition matrix are

$$\theta_{11} = m_1 I_1 (m_1 R)$$

$$\theta_{12} = X_1 (m_{qp} R)\{1 + i\kappa a\}$$

$$\theta_{13} = \frac{n}{R} K_n (m_{sh} R)$$

$$\theta_{14} = X_1 (m_{qs} R)\{i\kappa + b\}$$

$$\theta_{21} = \rho_f I_0 (m_1 R)$$

$$\theta_{22} = \left\{ Am_{qp}^2 (1 + i\kappa a) + i\kappa F \left( i\kappa - am_{qp}^2 \right) \right\} K_n (m_{qp} R) - \frac{2N}{R} X_3 (m_{qp} R)\{1 + i\kappa a\}$$

$$\theta_{23} = \frac{2nN}{R} X_2 (m_{sh} R)$$

$$\theta_{24} = \left\{ Am_{qs}^2 (i\kappa + b) + i\kappa F \left( i\kappa b - m_{qs}^2 \right) \right\} K_n (m_{qs} R) - \frac{2N}{R} X_3 (m_{qs} R)\{i\kappa + b\}$$

$$\theta_{32} = -\frac{2nN}{R} X_2 (m_{qp} R)\{1 + i\kappa a\}$$

$$\theta_{33} = -N \left\{ m_{sh}^2 K_n (m_{sh} R) - \frac{2}{R} X_3 (m_{sh} R) \right\}$$

$$\theta_{34} = -\frac{2Nn}{R} X_2 (m_{qs} R)\{i\kappa + b\}$$

$$\theta_{42} = L \left\{ 2i\kappa - a \left( \kappa^2 + m_{qp}^2 \right) \right\} X_1 (m_{qp} R)$$

$$\theta_{43} = \frac{i\kappa nL}{R} K_n (m_{sh} R)$$

$$\theta_{44} = L \left( 2i\kappa b - \kappa^2 - m_{qs}^2 \right) X_1 (m_{qs} R) \qquad\qquad A4.12$$

where the constants $A$, $C$, $F$, $L$, and $N$ are given by equation 1.9, and $a$ and $b$ are given by Equation 1.10.

# 5

# Synthetic Borehole Microseismograms, Rock Properties, and Waveform Log Interpretation in the Open Hole

Synthetic borehole microseismograms offer a powerful tool for the investigation of acoustic waveforms in boreholes. Even though the exact forward modeling of waveforms for a given borehole in a specified formation has proven difficult because of problems in modeling logging sources in sufficient detail, synthetic microseismograms are useful in understanding how rock properties affect waveforms in the open hole. In this chapter we explore the ways in which the seismic properties of rocks affect waveforms measured in boreholes, and how those effects can be separated from the effects of borehole diameter and logging tool configuration.

Borehole sources are assumed to be azimuthally symmetric pressure variations imposed at the surface of a logging tool of given radius and centralized in the borehole. The source spectrum is taken as a relatively smooth distribution of excitation energies across a specified frequency band. Early investigators found source deconvolution of acoustic waveform records obtained in boreholes to be unexpectedly difficult. This difficulty is assumed to be related to the great sensitivity of borehole microseismograms to the details of the source spectrum in the vicinity of cut-off frequencies for the critically refracted modes. The simple, broad-band sources used for synthetic microseismogram calculations in this chapter do not model known logging sources except for such general characteristics as bandwidth and centerband frequency. However, these synthetic microseismograms can be used to illustrate the relationship between rock properties and the characteristics of acoustic waveforms recorded in boreholes (Chen, 1982; Paillet and Cheng, 1986).

## 5.1 THE NATURAL RANGE OF SEISMIC VELOCITIES IN ROCKS

In the first chapter of this book it was noted that continuous elastic solids can be defined by two elastic constants and bulk density. These two elastic constants

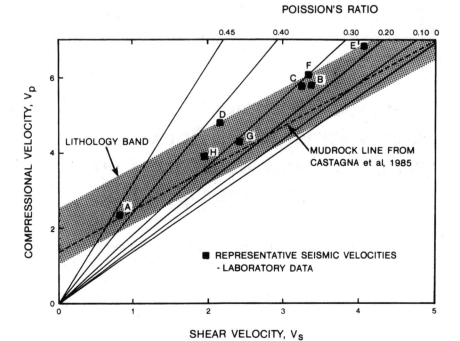

FIGURE 5.1.   Seismic velocities of typical sedimentary, igneous, and metamorphic rocks illustrating the lithology band in velocity space.

give rise to the propagation of two types of elastic body waves away from an arbitrary initial displacement. The velocities of these two waves, $V_p$ and $V_s$, are uniquely related to the elastic moduli of the homogeneous rock. Although the actual body wave velocities of elastic solids can vary over a wide range, seismic velocities of rocks are almost always confined within certain limits known as the "lithology band" in seismic velocity space (Domenico, 1977; Wilkens et al., 1984; Castagna et al., 1985; Han et al., 1986; Paillet et al., 1987a). The lithology band is illustrated in Figure 5.1. The Poisson's ratio of elastic solids is related to the elastic moduli and the ratio of compressional to shear velocities according to the equation (White, 1983)

$$\nu = \frac{V_p^{\,2} - 2V_s^{\,2}}{2\left(V_p^{\,2} - V_s^{\,2}\right)} = \frac{\gamma^2 - 2}{\left(2\gamma^2 - 1\right)}$$

$$\gamma = \frac{V_p}{V_s}$$

5.1

The softer, more plastic rocks such as shale and claystone have lower seismic velocities and relatively high values for the Poisson's ratio. For the purpose of a general description of seismic waves adjacent to boreholes, rocks may be divided into hard formations with both $V_p$ and $V_s > V_f$ and a Poisson's ratio of 0.20 to 0.30; and soft formations with $V_p > V_f > V_s$ and a Poisson's ratio of 0.30 to 0.45.

## 5.2 EFFECTS OF BOREHOLE DIAMETER AND MODE CONTENT ON SYNTHETIC MICROSEISMOGRAMS

One of the most evident facts in the interpretation of acoustic full waveform logs is the dependence of waveform appearance on borehole diameter even when rock properties are otherwise identical. The great difference in waveform appearance between waveforms obtained using the same source frequency range in different diameter boreholes stems from importance of borehole geometry in determining the numbers and character of trapped modes in the borehole. In this sense, the seismic properties of the rocks surrounding the borehole are almost secondary in importance to the borehole diameter and other geometrical factors such as wall smoothness and tool centralization.

An example of the effects of borehole diameter on borehole waveforms is given in Figure 5.2, where microseismograms are computed for a logging tool centralized in a fluid-filled borehole in sandstone. The relatively wide frequency band of the source extends from 0 to 20 kHz, with maximum source energy at 10 kHz. Rock and borehole fluid velocities and densities are given in the figure. The overall appearance of the waveforms changes significantly as the borehole diameter is increased, even though compressional and shear head waves arrive at nearly the same time in each microseismogram. The greatest difference in waveform appearance is associated with the dispersed pseudo-Rayleigh arrivals, which become larger in relative amplitude and more complicated in appearance as multiple modes are superimposed on the part of the waveform arriving after the head waves.

At the same time, the tube wave becomes less apparent in the full waveform. Some of the difference is caused by the usual practice of normalization in presenting both synthetic microseismogram and experimental waveform plots. That is, the plots are presented so that the largest wave amplitude just fills the allotted vertical scale. This is accomplished in the calculations by scaling time series plots so that maximum amplitude corresponds to a unit scale. Waveform log data are likewise scaled in the field using equipment gain controls so that maximum pressure fluctuations remain on scale without saturating. The significance of this gain control is illustrated in Figure 5.2. Calculations of tube-wave energy indicate that the tube wave is excited with about the same energy in all three examples. The tube wave only appears smaller in larger diameter boreholes because additional mode energy is carried by the pseudo-Rayleigh waves at larger borehole diameters.

The comparison of microseismograms calculated for different diameter boreholes using the same source in the same lithology and with the same borehole fluid appears to indicate that borehole effects will be a great problem in any attempts to interpret waveforms. However, the dependence of waveform

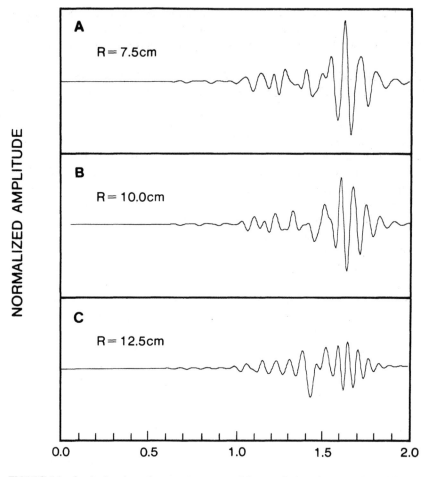

FIGURE 5.2. Synthetic microseismograms computed for a typical sandstone at three different borehole diameters, illustrating the effects of mode content on waveform appearance; sandstone lithology of Figure 4.7 with borehole diameter as given.

appearance on mode content indicates that there is a rough equivalence between frequency and borehole diameter. That is, the borehole will act as a similar waveguide whenever the same number of modes are excited at approximately the same ratio of energies. The examples given in Chapter 3 demonstrated that mode content of waveforms in borehole and channels is determined by constructive interference of waves traveling along the borehole or channel. Although the correspondence cannot be exact because of the differences in the radius of curvature of the borehole wall for different borehole diameters, waveforms obtained in different diameter boreholes will be similar if frequency and annulus width are scaled such that the same borehole modes are excited.

The approximate equivalence between borehole diameter and source frequency is illustrated in Figure 5.3. The figure compares microseismograms

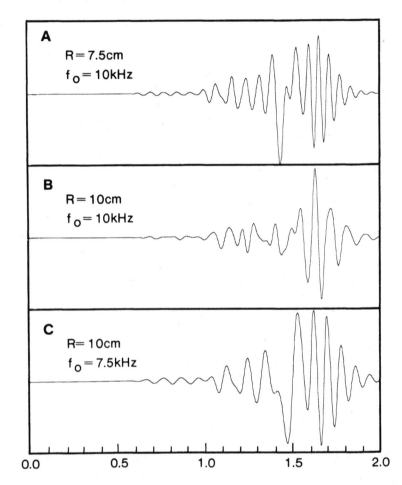

FIGURE 5.3. Comparison of synthetic microseismograms computed at two different borehole diameters with a third microseismogram scaled such that frequency compensates for increased diameter; the third microseismogram has the same mode content as the first microseismogram, but is computed for the same diameter as the second microseismogram; sandstone lithology of Figure 4.7.

computed at the same frequencies in different diameter boreholes with a third microseismogram computed for the larger diameter borehole, but at lower frequencies. For computation of the third waveform, source bandwidth and centerband frequency were scaled so that the ratio of wavelength to annulus width at centerband and band limit frequencies remained the same as those of the first waveform. This results in having the same number of modes with the same relative excitation in the first and third examples. The difference in borehole wall curvature introduces only minor differences in the computed microseismograms. This can be seen, for example, by noting the relative excitation of the compressional head wave in each of the three waveforms.

The importance of modes in determining waveform character in hard formations is illustrated in Figure 5.4 (Paillet, 1983b, 1984). The mode composition of three recorded waveforms recorded in the field under three different conditions in the same granite formation illustrate how the superposition of modes determines the appearance of the composite waveform. In the case of excitation at frequencies below the lowest pseudo-Rayleigh and PL mode cut-off frequencies, the waveform is completely dominated by the tube wave (Figure 5.4C). If waveform amplitude is scaled by normalizing with respect to the maximum amplitude, the head waves cannot be identified in the resulting plot. When borehole gains are increased and digitizing intervals decreased, the waveforms indicate that head waves are excited by the high-frequency edge of the source spectrum (Figure 5.4D). If frequency or borehole diameter are increased so that centerband frequency corresponds to the cutoff frequency for the lowest pseudo-Rayleigh and compressional modes, head waves are excited with easily detected amplitudes. The waveform in Figure 5.4B represents the optimum conditions for head wave generation as described in the previous chapter.

The top waveform in Figure 5.4 (Figure 5.4A) illustrates the superposition of multiple modes in the case where borehole diameter is large enough or frequencies high enough to allow several pseudo-Rayleigh modes to propagate. The composite waveform is severely complicated by the interference pattern produced by the three lowest modes. A similar interference pattern characterizes the compressional arrivals. The "beating" of PL modes produces the repeating series of compressional arrivals noted by Koerperich (1979, 1980). These multiple arrivals have previously been attributed to differences in fluid delay associated with multiple refractions. The complicated and unpredictable changes in phase between receiver stations caused by the superposition of multiple modes severely complicate the picking of shear velocity from waveforms in the field (Paillet and Cheng, 1986). These results explain the difficulty encountered in attempting to determine shear transit times from waveform logs obtained in large diameter boreholes.

## 5.3    SYNTHETIC MICROSEISMOGRAMS IN HARD FORMATIONS — EFFECTS OF LITHOLOGY

The synthetic microseismogram theory presented in the previous chapter indicates the importance of mode content in determining the appearance of waveforms. Compressional and shear travel time are determined by the first head wave arrivals, which depend upon the properties of the leaky PR modes listed in Table 3.1. Other characteristics of waveforms that depend upon seismic velocities are more difficult to recognize. The only other important control on waveform appearance for hard formations is the Poisson's ratio. Cheng and Toksöz (1981) demonstrated that the Poisson's ratio of formations influences the amplitude of the waveform in the time window between compressional and shear head wave arrivals (Figure 5.3). Their results show that the amplitude in this window increases as the Poisson's ratio is increased, corresponding to a decrease in $V_s$ if $V_p$ is held constant. This amplitude increase is associated with increased

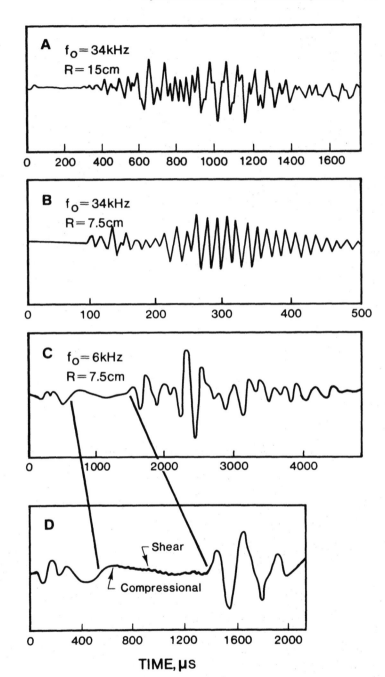

FIGURE 5.4.  Waveforms obtained in boreholes in homogeneous granite illustrating frequency scaling effects: (A) waveform corresponding to superposition of three pseudo-Rayleigh modes, (B) waveform corresponding to source frequency near cutoff for the lowest pseudo-Rayleigh mode, (C) waveform corresponding to source frequency below all mode cutoffs, and (D) waveform illustrating weakness of compressional and shear head waves. (These wave arrivals are associated with excitation by the high frequency edge of a low frequency source spectrum).

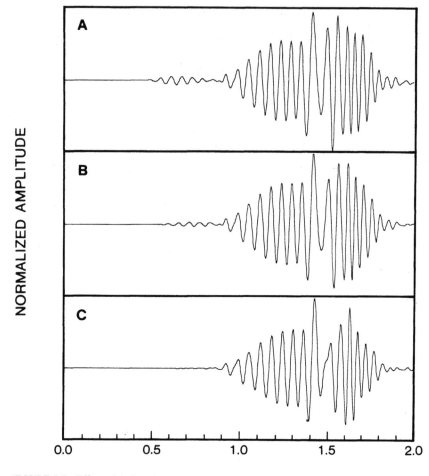

FIGURE 5.5. Effect of Poisson's ratio on PL mode excitation; synthetic microseismograms calculated for increasing $V_p$ such that Poisson's ratio equals (A) 0.1, (B) 0.26, and (C) 0.32; fluid is water, $R = 15$ cm, $V_s = 2.3$ km/s, and $\rho_s = 2.3$ g/cm³.

excitation of the PL mode. Microseismograms in Figure 5.5 compare waveform computations when $V_s$ is held constant, and $V_p$ is increased, corresponding to decreasing Poisson's ratio. The amplitude of the PL mode (wave energy between compressional and shear arrivals) decreases by a factor of about 10 as the Poisson's ratio decreases from 0.32 to 0.10. The coupling of source energy to the PL mode appears inversely related to the ratio $V_p/V_f$ when $V_s > V_f$.

Another important consideration in the character of waveforms in hard formations is attenuation. The attenuation of waveforms in boreholes is determined by the combined compressional and shear attenuation of the formation and the attenuation of the borehole fluid. The dependence of each wave mode on these three different sources of attenuation is determined by the partition coefficients introduced in the previous chapter. If waveforms are plotted in the usual normalized format, the differential attenuation of the modes produces a substantial change in waveform appearance. The generally lower attenuation for the compressional mode enhances the compressional arrival, while the pseudo-Rayleigh arrival is decreased.

## 5.4    SYNTHETIC MICROSEISMOGRAMS IN SOFT FORMATIONS

The appearance of synthetic microseismograms becomes completely different from that for hard formations illustrated in Figures 5.2 to 5.4 when $V_s$ becomes less than $V_f$. In that case, the pseudo-Rayleigh modes are not generated, and most wave energy is coupled into compressional normal modes (Table 3.1). The compressional modes are similar to the PL modes in the hard formation case except that mode attenuation by outward shear radiation is not great, and excitation energies of modes can be significant. An example of synthetic microseismograms for a fluid-filled borehole in a soft formation is given in Figure 5.6. The first part of the figure shows the dispersion curves for the compressional modes. They appear similar to the pseudo-Rayleigh dispersion curves in hard formations except that phase velocity equals $V_p$ at mode cutoff. These modes also have at least a small amount of attenuation because outward radiation losses through shear wave propagation in the rock are possible. The synthetic microseismograms are dominated by the airy phase of the compressional modes in the same way that maximum amplitude in the pseudo-Rayleigh modes was associated with the minimum in the group velocity curve.

One of the major differences between synthetic microseismograms in hard and soft formations is the appearance of the tube wave (Figure 5.7). The tube wave appears to travel at a softer velocity and to contain much lower frequencies for the soft formations. The excitation amplitudes shown in the first part of the figure explain the difference. The source frequency band excites tube waves at all frequencies, although the amplitude is shown to level off for frequencies greater than about 10 kHz. Synthetic microseismograms computed using source frequencies greater than 10 kHz contain the tube wave, although it is difficult to recognize because pseudo-Rayleigh waves are excited with even greater amplitudes. In contrast, the excitation function for tube waves in soft formations falls off rapidly for frequencies greater than 3 kHz. This phenomenon explains the once confusing irregularity with which tube waves are identified in waveform logs. Conventional acoustic logging sources are designed with source centerband frequency of at least 10 kHz. In soft formations, tube-wave excitation would therefore depend upon the lower edge of the source frequency band. Minor differences in the source amplitude at these frequencies would make a great difference in the amplitude of the excited tube wave. For this reason, apparently similar sources or slightly different signal processing filters could make a much greater difference in tube-wave appearance than would be expected. Otherwise, all of the statements about the relationship between mode content and waveform characteristics made for the hard formation waveforms apply to the waveforms

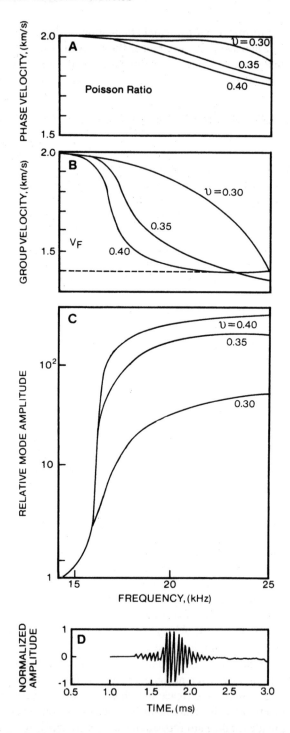

FIGURE 5.6.   Synthetic microseismogram computation for "slow" formation: (A) phase velocity of lowest PL mode, (B) group velocity of lowest PL mode, (C) mode excitation for lowest PL mode, and (D) synthetic microseismogram for Poisson's ratio = 0.40; fluid is water.

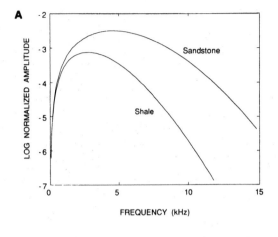

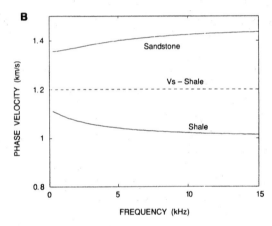

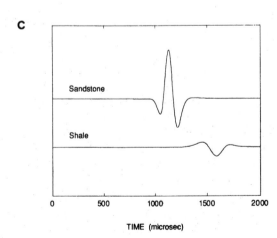

FIGURE 5.7.   Comparison of tube-wave excitation in "fast" and "slow" formations: (A) tube-wave excitation, (B) tube-wave phase velocity, and (C) synthetic microseismograms for tube waves.

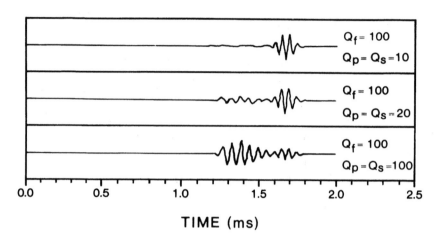

FIGURE 5.8.   Effects of attenuation on synthetic microseismograms computed for "slow" formations — microseismograms computed for three different sets of $Q_p$ and $Q_s$; fluid is water, $V_p$ = 2.0
km/s, $V_s$ = 2.0 km/s, and $\rho_s$ = 2.0 g/cm³.

in soft formations, except that compressional normal modes rather than pseudo-
Rayleigh modes are dominant at typical logging frequencies.

The effect of attenuation on synthetic microseismograms in soft formations
is illustrated in Figure 5.8 (Cheng et al., 1986; Toksöz et al., 1985). Waveforms
computed for a typical shale formation are presented for various $Q_p$ and $Q_s$
values. The usual normalization produces the expected increase in amplitude of
the compressional head wave relative to the later-arriving, higher amplitude
parts of the waveform. As in the case of the hard formations, the increase is
attributed to the greater dependence of the higher-frequency parts of the mode
excitation on fluid attenuation.

## 5.5   COMPARISON OF SYNTHETIC MICROSEISMOGRAMS FOR
THE NATURAL RANGE OF LITHOLOGIES

Many examples of synthetic microseismograms have been used to illustrate
the large effects of a borehole diameter and mode content on borehole waveforms. These effects appear to be the most important consideration in the analysis
of specific microseismograms computed for specific situations. However,
Paillet and Cheng (1986) demonstrated that the most effective conditions for
excitation of head waves in boreholes correspond to source frequency band
centered on the cutoff frequencies for the lowest compressional and shear modes.
Almost all borehole logging systems have been designed to correspond to these
conditions. One can compare microseismograms calculated for a series of points
along the lithology band in Figure 5.1 where the source frequency band is
centered on the lowest pseudo-Rayleigh cutoff frequency when $V_s > V_f$, and for
the case where the source frequency band is centered on the lowest PL mode
cutoff frequency when $V_s < V_f$. Microseismograms were calculated at eight

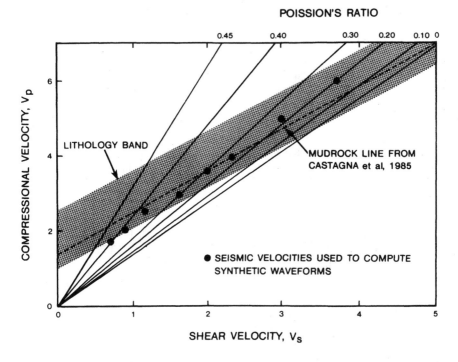

FIGURE 5.9.    Eight points selected for computation of synthetic microseismograms superimposed on the lithology band of Figure 5.1.

points along the lithology band (Figure 5.9) for lithologies ranging from crystalline basalt to homogeneous shale. The microseismograms are plotted in Figure 5.10 using the usual normalization. On this scale of representation, the shear head wave arrivals are easily recognized for the first three waveforms where $V_s > V_f$. The waveforms become dominated by compressional modes when $V_s < V_f$, and the tube wave begins to lag far behind the rest of the waveform energy.

The waveforms in Figure 5.11 do not include the effects of attenuation. Figure 5.11 illustrates the synthetic microseismograms that result using values of attenuation for the appropriate rock type, and with the same attenuation, but normalized to the maximum amplitude in each waveform. The greater attenuation and more incomplete refraction for the softer formations produce a great decrease in amplitude with decreasing seismic velocities. The normalization of waveforms emphasizes the transition to compressional modes instead of pseudo-Rayleigh modes and the trend towards decreasing velocity and frequency of the tube wave for decreasing seismic velocities. In cases where waveform logs are obtained in the field, recognition of lithologies can be complicated by the simultaneous changes in waveform amplitude and character. If gains are kept constant over the entire interval, soft formations will be characterized by greatly reduced amplitudes. If waveform gains are allowed to vary in such a way that

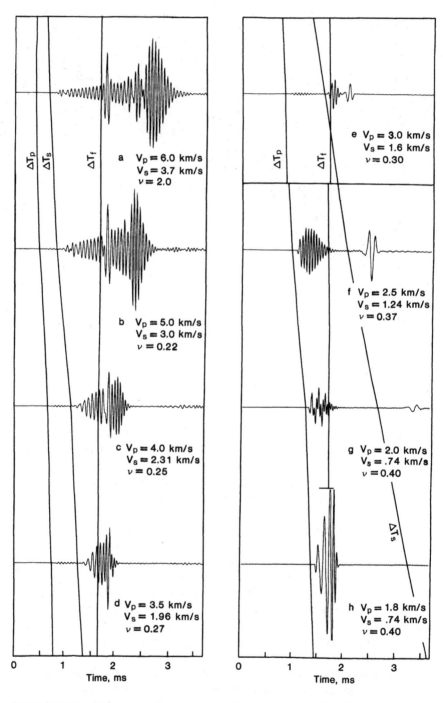

FIGURE 5.10. Synthetic microseismograms (normalized and unattenuated) for the eight points in velocity space illustrated in Figure 5.9.

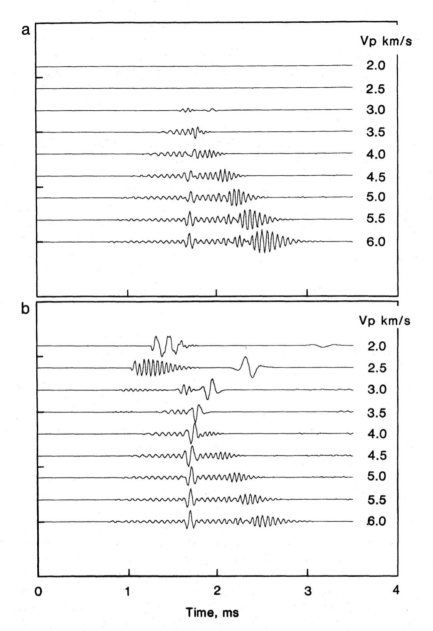

FIGURE 5.11. Synthetic microseismograms computed for the eight points in Figure 5.09 with (A) typical attenuation values and (B) attenuated microseismograms with normalization used in waveform logging.

each individual waveform is normalized on the basis of the largest amplitude in the time series, changes in waveform character may give a very different impression of lithologies. For example, comparison of the two sets of waveforms in Figure 5.11 indicates that the shift of mode excitation from pseudo-Rayleigh to PL modes as $V_s$ decreases can make the energy appear to travel faster in soft formations. Perhaps the best indicator of the difference between the normalized waveforms in Figure 5.11B is the consistent decrease in apparent frequency of the tube wave with decreasing compressional velocity and increasing Poisson's ratio.

## 5.6    COMPARISON OF SYNTHETIC MICROSEISMOGRAMS WITH BOREHOLE WAVEFORMS

Synthetic borehole microseismograms provide a useful means for investigating the effects of borehole conditions and seismic velocities on acoustic waveforms. However, the insight into the generation of head waves and other modes obtained from such calculations indicate the great sensitivity of borehole wave propagation to source frequency content. To what extent can synthetic borehole microseismograms be used to compute waveforms which closely correspond to waveforms obtained in boreholes? Our experiences indicate that almost all of the characteristics of experimental waveforms can be reproduced in the computations after several iterations of revised estimates for seismic velocities, fluid properties, and attenuation. In many situations this information comes from core samples where differences in *in situ* conditions during testing, core property alteration during laboratory processing of samples, and drilling damage to the borehole wall need to be taken into account when comparing borehole wave forms with computed microseismograms. After considering all of the uncertainties in the parameters required to model wave propagation in the borehole, the synthetic microseismogram computations appear to account for all of the properties of measured waveforms, even if the precise details of specific data sets are very difficult to match.

An example of the degree of correspondence between computed waveforms and experimental data is illustrated in Figure 5.12. The figure compares synthetic microseismograms for two formations with waveforms obtained in boreholes in limestone and a shaley sandstone. The first of these clearly qualifies as a hard formation, while the second falls into an intermediate region on the lithology band in Figure 5.1. Synthetic microseismograms were modeled using a decaying exponential source (Tsang and Rader, 1979) representing borehole forcing by a typical magnetostrictive logging source at a resonant frequency of 15 kHz. Normalized synthetic microseismograms computed without attenuation do not closely correspond with the experimental data. However, when attenuation is included in the calculations, the correspondence between synthetic microseismograms and experimental waveforms is greatly improved. In general, the correspondence between borehole waveforms and the synthetic seismograms indicate that the theory accounts for all of the physical wave propagation processes that determine the character of the observed waveforms.

One of the most significant results obtained from the synthetic microseismogram calculations is the effect of borehole diameter on mode

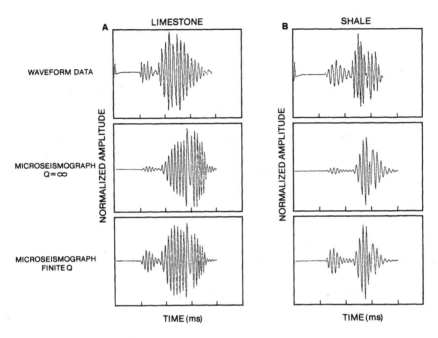

FIGURE 5.12. Comparison of synthetic microseismograms (with and without attenuation); actual waveform data obtained in (A) limestone and (B) shale.

excitation and waveform characteristics. A comparison between the borehole waveforms obtained in a granite formation at different borehole diameters was given in Figure 5.4. The mode calculations indicate that the second waveform (Figure 5.4B) corresponds to the optimum conditions for borehole logging. That is, source centerband frequency (34 kHz) is close to the low frequency cutoffs for the lowest PL and pseudo-Rayleigh modes. The waveform contains well-defined compressional and shear head wave arrivals, while the subsequent part of the pressure series is dominated by the tube wave. A waveform obtained in the same formation using the same source frequency in a larger diameter borehole is illustrated in Figure 5.4C. This waveform contains the interference pattern resulting from the superposition of at least two different PL and pseudo-Rayleigh modes. Compressional and shear head wave arrivals appear much less regular than in the first figure.

Another test of the synthetic microseismogram theory is given in Figure 5.4A, where waveforms are given for source centerband frequency far below cutoff for all PL and pseudo-Rayleigh modes in the same granite. The experimental waveform obtained with the usual gain control settings contains only the tube wave. However, the same interval was relogged with all logging system gains raised to the limit. In that case, the amplitude on the tube wave is completely saturated in the recordings. The beginning of this experimental waveform is also illustrated in Figure 5.4D, indicating a train of low-amplitude arrivals that appear to be the shear head wave. The frequency of these arrivals is hard to define because of the low amplitudes, but appears to be close to the 25 kHz frequency for the lowest pseudo-Rayleigh mode cutoff determined for the borehole diameter and seismic velocities given in the figure. A number of these multiple-frequency borehole waveform experiments confirm the usefulness of the synthetic microseismogram theory in predicting the characteristics of borehole waveforms in formations of known seismic velocities and attenuation (Paillet, 1983b; Paillet and Cheng, 1986). In those cases where seismic velocities are not well known, synthetic microseismograms and a knowledge of how borehole conditions and rock properties influence waveform appearance can be used to converge on an acceptable match between synthetic microseismograms and experimental waveforms.

Much of the early application of acoustic waveform logging was for the investigation of the porosity, permeability, and extent of fracturing in hard formations. The synthetic borehole microseismogram calculations predict some very substantial differences from the typical fast formation results in the case of soft formations. The most important differences include the generation of large-amplitude compressional or PL modes instead of pseudo-Rayleigh modes, and the generation of tube waves with much lower frequency content if enough source energy exists at frequencies less than 3 kHz. An example of the correspondence between waveforms and synthetic microseismograms for a soft formation is given in Figure 5.13. The figure compares the borehole waveform with a synthetic microseismogram for ocean floor sediments where seismic velocities are reconstructed from surface seismic data and core samples. The calculations indicate that the large-amplitude arrivals in the experimental data correspond to the compressional mode. If a tube wave had been present, the frequency content and slow velocity of the tube wave would be easy to recognize. The correspondence between data and calculations otherwise demonstrate the extent to which the theory predicts mode content of waveforms in soft formations. However, the iteration process involved in matching synthetic microseismograms to waveforms becomes more difficult for slow formations where undisturbed core samples are difficult to obtain, and samples are subject to desiccation and other forms of alteration during processing. Our results generally confirm that waveforms can be modeled with considerable accuracy if enough data is available.

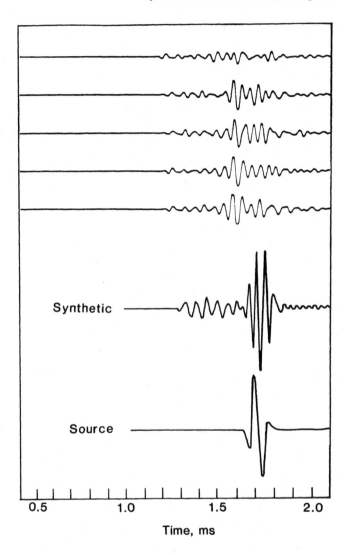

FIGURE 5.13. Comparison of synthetic microseismogram with actual waveform data for soft seabed sediments (from Toksöz et al., 1985).

# 6

# Acoustic Log Interpretation Methods in the Open Hole

In this chapter we discuss methods of picking formation velocities and attenuation from full waveforms, and the interpretation of the picked velocities and attenuation for porosity and lithology. In conventional sonic logging, velocity picking is usually achieved by threshold detection at two different receiver distances, and the moveout in time between the two picks divided by the distance between the receivers gives the P-wave transit time of the formation (Figure 1.1). In this chapter we are interested in the picking of formation velocity and attenuation from full waveform logs. Because waveform frequencies and amplitudes decrease with source-receiver spacing, threshold picking becomes less reliable and less accurate. In the following sections we will first discuss the various arrivals in a full waveform log microseismogram, and then present methods of determining formation properties from waveform log data. We will be especially concerned with the identification of how such variables as receiver spacing, frequency content, etc., affect the quality of the interpretation.

## 6.1 FREQUENCY RESPONSE IN BOREHOLES: MODE CONTENT AND FIRST ARRIVALS

As discussed previously in this book, the content of the wave train propagating along a borehole is made up of several different modes, including guided waves and head waves. The first arrival in any full waveform acoustic log wave train is the compressional (or P) head wave. This arrival is immediately followed by the P leaky (PL) modes. Depending on the frequency of excitation and the radius of the borehole, there may be one or more of the P leaky modes. If the frequency band of the exciting source is narrow, and falls within the frequency range of one of the P leaky modes, then a relatively long and ringy P wave train will result. Sometimes this wave train will carry into the shear/pseudo-Rayleigh wave train and interfere with the proper identification of the latter. This is especially true at shorter source-receiver spacing and/or at lower frequencies.

After the P wave train, in a fast formation, the next arrival is the S head wave, followed immediately by the pseudo-Rayleigh wave. Analogous to the P leaky modes, the pseudo-Rayleigh wave has an infinite number of modes and none, or one or more of them, could be excited, depending on the source characteristics.

In the normal logging frequency band, the pseudo-Rayleigh mode is usually of larger amplitude than the P wave train. This is because the pseudo-Rayleigh mode is a guided wave in the borehole, with no geometric spreading or attenuation. In contrast, the P wave train in constantly radiating energy away from the borehole as shear waves, hence the name P leaky mode. Analyses have shown that in the far field the P wave train has a geometric spreading factor inversely proportional to the source-to-receiver separation. At far offsets, the pseudo-Rayleigh wave will have a larger amplitude than the P wave. In cases with a low frequency source or a small borehole, the pseudo-Rayleigh wave is poorly excited and may not be detectable under normal gain settings. In a slow formation, where the shear wave velocity is lower than the acoustic velocity of the borehole fluid, no S head wave or pseudo-Rayleigh wave exist.

The tube wave arrival is usually quite easy to identify. In an iso-offset section, the tube wave appears to be straight, parallel lines, with little variations with lithology. This is because the tube wave properties are largely controlled by that of the borehole fluid. The tube wave arrival is usually of lower frequency content than either the P or the S or pseudo-Rayleigh waves. This is because the excitation of the tube wave (before multiplication by the factor $\rho\omega^2$ in Equation 4.13) is almost inversely proportional to frequency. When excited with a broad-band source, the low frequency character of the tube wave is very evident. In a slow formation, there is no S or pseudo-Rayleigh arrival, as the tube wave appears as the next prominent arrival in the full waveform record. Field data examples of full waveform logs in a fast and a slow formation are shown in Figures 6.1 and 6.2.

## 6.2  PICKING COMPRESSIONAL VELOCITY: EXAMPLES WITH SYNTHETIC MICROSEISMOGRAMS

The compressional or P wave is the first arrival in a microseismogram. It is thus easy to identify. Conventional methods for picking the compressional wave arrivals from two or more receivers and then measuring their moveout consist of picking the first peak, first trough, or the first zero crossing. All three methods are quite robust. However, because full waveform acoustic logs are usually digitized at 5 μsec intervals and have receiver-receiver separations of about 30 cm, the accuracy involved in these techniques is no better than one digitization interval divided by the offset. With the typical numbers quoted here, that amounts to 16.7 μsec/m. In a hard limestone environment, with typical P wave transit time of about 167 to 200 μsec/m, this amounts to an unacceptable 10% error. In the following, we will demonstrate all three different picking schemes with two synthetic microseismogram examples, along with inherent limits on the accuracy of these methods.

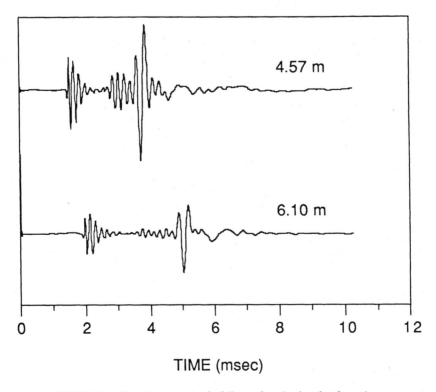

TIME (msec)

FIGURE 6.1     Field data example of a full waveform log in a fast formation.

## 6.2.1.     Velocity Picks in Two Formations Using Synthetic Examples

The first example is for a sandstone formation, with $V_p = 4.3$ km/s and source to receiver separations of 1.5 and 2 m. The center frequency is taken to be 10 kHz. Figure 6.3 shows the synthetic waveforms and the corresponding picks using (a) first peak, (b) first trough, and (c) first zero crossing. The corresponding velocities are (a) 4.27 km/s, (b) 4.27 km/s, and (c) 4.34 km/s. This is achieved with a sampling interval of 1.95 μsec and a receiver-receiver separation of 0.5 m, giving a transit-time accuracy of 3.9 μsec/m. Given the known formation transit time of 232.6 μsec/m, this error is slightly less than 2%. The difference in the picked velocities results from one sample point difference. Because of the digitization, it was impossible in this case to obtain the exact formation velocity. The two velocities picked fall on either side of the exact answer.

Figure 6.4 shows the same picks for a shaley formation, with $V_p = 2.5$ km/s, equal to a transit time of 400 μsec/m. In this case the sampling accuracy of 3.9 μsec/m results in an error of less than 1%. Velocities determined using (a) the

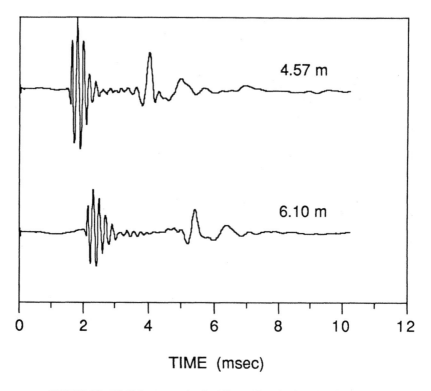

FIGURE 6.2    Field data example of a full waveform log in a slow formation.

first peak, (b) the first trough, and (c) the first zero crossing are 2.46, 2.49, and 2.51 km/s, respectively. The differences in the velocities represent two digitization points. Once again, the correct formation velocity was impossible to obtain due to digitization. However, the relative error is much smaller than the sandstone case because of the larger travel time. In both these examples, the digitization intervals are smaller than those used in actual logging, and the receiver to receiver separation longer; thus picking errors are smaller than one would encounter in actual data.

One way to obtain better accuracy in velocity picking from digitized waveforms is to cross-correlate the two P wave waveforms, and then interpolate the cross-correlation function to obtain the moveout to an accuracy better than the digitization interval (Cheng et al., 1981). We can safely do this because we have over-sampled the waveform to begin with. With a 5 μsec digitization interval we are dealing with a Nyquist frequency of 100 kHz, while the signal is concentrated at around 5 to 15 kHz. Interpolating between the digitized time points is equivalent to adding zero to the high frequency end of the spectrum and thus will not be affecting the signal itself directly. In theory, one can get as fine a resolution as one desires. In practice, an improvement in accuracy of an order of magnitude is easily obtainable. A simple interpolation algorithm involves fitting the cross-correlation coefficients with a polynomial (preferably a Cheybeshev polyno-

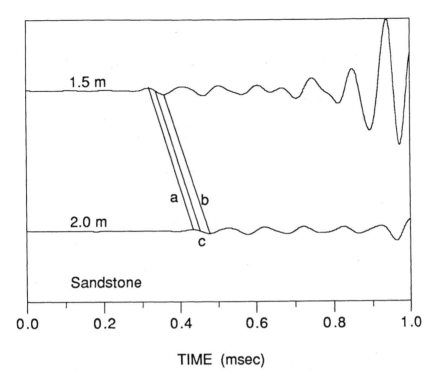

FIGURE 6.3.  Compressional velocity picks based on (a) first peak, (b) first zero crossing, and (c) first trough for typical sandstone example with $V_p = 4.3$ km/s and source-to-receiver separations of 1.5 and 2.0 m.

mial), then evaluating the polynomial at several (say 10) points between the two highest cross-correlation coefficients, and taking the maximum. There are several other variations based on this idea; some of them are given in Willis and Toksöz (1983).

## 6.2.2    Depth of Penetration, Borehole Wall Alteration, and Long-Spaced Sonic Measurements

One of the more frequently asked questions is what is the depth of penetration or investigation of the full waveform logging tool. In an unaltered formation, the depth of investigation is generally taken to be about one wavelength of the P wave, which is of the order of 50 cm. However, in a borehole with alteration in the borehole wall, the answer is more complex. If the damaged or altered zone around the borehole results in a velocity gradient increasing away from the borehole, then the seismic energy will be refocused back into the borehole. The depth of investigation then depends on the velocity gradient at the borehole wall, as well as the source-receiver separation. Stephen et al. (1985) modeled this damaged zone case using the finite difference method. Their results are shown in Figure 6.5. They found that ray tracing results can be used to predict the depth

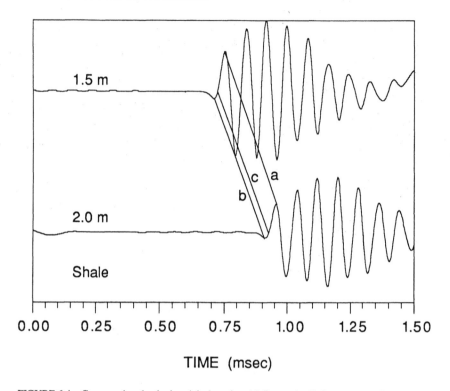

FIGURE 6.4.  Compressional velocity picks based on (a) first peak, (b) first zero crossing, and (c) first trough for typical shale example with $V_p$ = 2.5 km/s and source-to-receiver separations of 1.5 and 2.0 m.

of investigation and the minimum source-receiver separation to see past a damaged zone, despite the relatively low frequency and long wavelength involved. These calculations confirm the application of standard "cross-over" time refraction formulae (Grant and West, 1960) given by Baker (1984) as described in Chapter 2.

## 6.3    PICKING SHEAR WAVE VELOCITY

One major advantage in use of the full waveform acoustic log over the conventional sonic log is the ability to determine from it the formation shear wave velocity. Shear wave velocity provides additional important information in formation evaluation. For example, the ratio of compressional to shear wave velocity, or equivalently, the Poisson's ratio, is a good indicator of lithology and gas saturation. The shear modulus of a formation can be calculated using the shear wave velocity and density (from gamma density log) and is very important in designing a hydrofrac experiment. Lower shear wave velocity and amplitude could be an indicator of *in situ* fractures.

The determination of shear wave velocity from full waveform acoustic logs

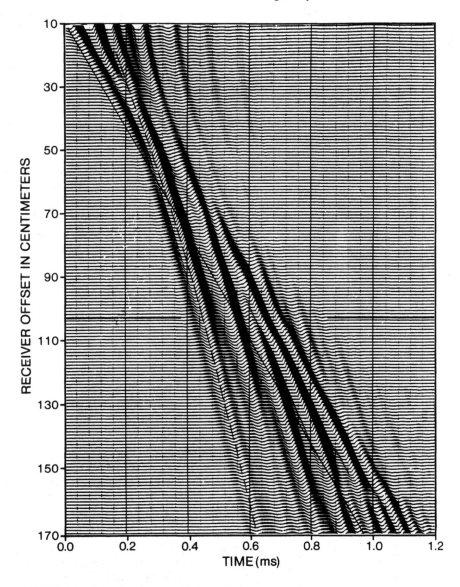

FIGURE 6.5.    Finite difference model of waveforms propagating along a borehole in sandstone with damaged zone modeled by an annular layer 10 cm thick and characterized by a compressional velocity decrease of about 38%; first arrivals refracted from the unaltered formation first appear at about 102 cm; from Stephen et al. (1985).

is a much more complicated problem than picking the P wave velocity. Unlike the P wave, which exists in all formations, the shear wave, which is a refracted wave, exists only in a fast formation where the shear wave velocity is higher than the acoustic velocity of the borehole fluid (drilling mud or water). In a slow formation, where the shear wave velocity is lower than that of the acoustic velocity of the borehole fluid, there is no refracted shear wave. As a result, in these formations we have to use indirect techniques to determine the formation shear wave velocity. Some of these techniques are described later in this chapter.

Even in a fast formation, identifying the shear wave in the full waveform seismogram and then measuring the moveout is not a straightforward process. This is because of the existence of the pseudo-Rayleigh wave, which travels at the formation shear wave velocity at its low frequency cutoff and which has no geometric dispersion, resulting in a much larger amplitude than that of the refracted shear wave. The superposition of the pseudo-Rayleigh wave modes on the shear wave arrival makes the detection and picking of the shear wave more difficult than that of the compressional wave. On the other hand, because the pseudo-Rayleigh wave travels at the shear wave velocity at cutoff and has no geometric dispersion, it usually appears as a large amplitude event starting at the formation shear wave arrival time. Thus, although we may be picking the pseudo-Rayleigh instead of the formation refracted shear wave, the moveout is still at formation shear wave velocity.

### 6.3.1    Shear Velocity from Waveforms ("Visual" Picks); Example From a Fast Formation

The first example is from the sandstone synthetic microseismograms shown in Figure 6.3. The formation shear wave velocity is 2.0 km/s. It is obvious where the large shear/pseudo-Rayleigh wave packet begins. Willis and Toksöz (1983) used an algorithm that identifies the shear/pseudo-Rayleigh wave arrival as the first arrival after the P wave arrival traveling at least 1.4 times the P wave travel time. The latter condition is to guarantee that the Poisson's ratio is nonnegative. The former condition takes advantage of the large amplitude of the pseudo-Rayleigh wave. This algorithm fails when the borehole diameter is so small that the pseudo-Rayleigh cutoff frequency is pushed outside the power band of the source. In these cases shear wave arrival picking will have to be done on a trace-by-trace basis.

In our synthetic example shown in Figure 6.6, picking (a) the first peak and (b) the first trough of the shear/pseudo-Rayleigh wave packet gives us velocity of 1.98 and 1.95 km/s, respectively. These velocities are slightly lower than the formation shear wave velocity. This is common for shear wave picks because of the interference and dispersion of the pseudo-Rayleigh wave mentioned earlier.

### 6.3.2    Semblance Cross-Correlation and Other Advanced Shear Picking Methods

As in the case for the compressional wave, cross-correlation helps in improving the accuracy of the shear wave velocity determined. However, owing to the emergent nature of the shear/pseudo-Rayleigh packet, there are sometimes doubts about whether the pick is the first arrival of the shear wave. In order to cut down the influence of one bad pick and improve the accuracy of the moveout, Willis and Toksöz (1983) used a semblance cross-correlation algorithm. A

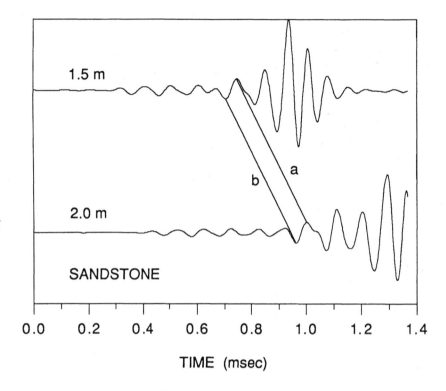

FIGURE 6.6.   Shear velocity picks based on (a) first peak and (b) first trough for typical sandstone example with $V_s$ = 2.0 km/s and source-to-receiver separations of 1.5 and 2.0 m.

schematic diagram for the algorithm is shown in Figure 6.7. Semblance cross-correlation takes into account the relative amplitudes as well as the coherency of the waveforms. In the Willis and Toksöz algorithm, the compressional wave is first extracted from the near receiver waveform, is correlated along the rest of the waveform and the shear wave delay is then picked as that correlating lag which has the larger semblance correlation value past 1.4 times the P wave arrival time. The shear/pseudo-Rayleigh wave packet is then extracted and semblance cross-correlated with the far receiver starting with the known lag to refine the velocity pick. Willis and Toksöz claims that this is the most robust and accurate of several cross-correlation methods they have tested.

## 6.3.3    Mode Content and Shear Wave Picking

Because of the interference of the pseudo-Rayleigh wave with the refracted shear head wave, the frequency range of the exciting source plays a very important role in the accuracy and consistency of shear wave picking. If the frequency of excitation is broad enough to include more than one pseudo-Rayleigh mode, then shear arrivals become complicated. Since the fundamental and higher modes of the pseudo-Rayleigh wave all start at the formation shear

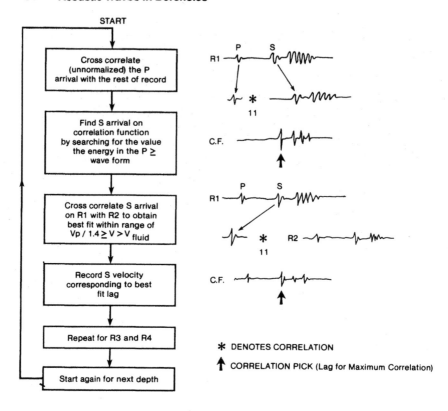

FIGURE 6.7. Schematic diagram illustrating the semblance cross-correlation algorithm used to pick shear velocities from acoustic waveform log data.

wave velocity and have dispersion curves that are similar in shape but offset in frequency, an excitation of two or more modes of the pseudo-Rayleigh wave results in the beating between the two modes. Furthermore, the high frequencies in a given acoustic source spectrum attenuate faster than low frequencies over a fixed interval because of the constant $Q$ assumption. Therefore, the relative amplitudes of the pseudo-Rayleigh modes will change with offset. The beating and hence the waveform will not be consistent from offset to offset, making the proper identification and picking of shear wave velocity very difficult. An example of this difficulty is shown in Figure 6.8 with actual field data.

### 6.3.4    Shear Logging with Shear Sources

The most direct method of measuring the shear wave velocity of the formation is direct shear wave logging. This is achieved by using nonaxisymmetric sources. Currently used nonaxisymmetric sources can be divided into two types: dipole sources and quadrupole sources. A dipole source is generated with a positive displacement of the borehole fluid in one direction and an equal but negative displacement 180° away in azimuth. This is usually accomplished using a bender element (White, 1967; Zemanek et al., 1984) or a movable cylinder (Kitsunezaki,

| NEAR RECEIVER PICK | FAR RECEIVER PICK | EQUIVALENT SHEAR VELOCITY, in km/s |
|---|---|---|
| A. LOW FREQUENCY TRANSDUCER ||| 
| a | a | 6.10 |
| b | b | 3.39 |
| c | c | 3.35 |
| d | d | 3.35 |
| e | e | 3.31 |
| f | f | 3.31 |
| B. HIGH FREQUENCY TRANSDUCER |||
| a | a | 5.26 |
| b | b | 5.08 |
| c | c | 2.88 |
| d | d | 4.23 |
| e | e | 2.63 |
| f | f | 2.93 |

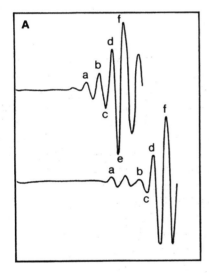
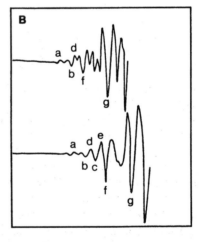

FIGURE 6.8. Example of shear velocity picking using various combinations of peak-to-peak and trough-to-trough picks; the interference pattern produced by the superposition of multiple modes causes the various velocity picks to vary substantially from the known shear velocity of 3.35 km/s.

1980; Kaneko et al., 1989). The resulting radiation pattern of the pressure generated in the borehole has a cosine $\theta$ dependence, where $\theta$ is the azimuthal angle. A quadrupole source is generated with two positive displacements 180° from each other and 90° from the positive displacements. The resulting radiation pattern of the source pressure field has a cosine $2\theta$ dependence. Chen (1988) gives a good description of this type of shear wave logging source.

The theory behind shear wave logging using dipole and quadrupole sources is more complex than the theory of acoustic logging using axisymmetric sources described in this book so far. The properties of nonaxisymmetric acoustic waveforms in boreholes are illustrated using synthetic microseismograms in Chapter 4. Interested readers are referred to Schmitt (1988b) for a more complete overview of the theory involved. Here we will discuss some of the results obtained from the theory which have important applications in shear wave logging.

The wave modes excited by either a dipole or quadrupole source in a fluid-filled borehole are guided waves that are analogous to the tube or pseudo-Rayleigh waves excited by an axisymmetric or monopole source. The wave modes generated by a dipole source are known as flexural or bending modes; those generated by a quadrupole source are known as screw modes. The fundamental or the lowest order flexural and screw modes are analogous to the tube or Stoneley waves, that is, they are interface waves resulting from the existence of a liquid-solid interface. The higher order flexural and screw modes are analogous to pseudo-Rayleigh modes in that they require the existence of a waveguide (the borehole). As such, these higher order modes cannot exist in slow formations since there is no shear wave critical angle, refraction, and postcritical reflection in these formations.

Unlike the case with the monopole source, both the fundamental and the higher order modes propagate with the formation shear wave velocity at the low frequency cutoffs. In case of the fundamental modes, since they are analogous to the tube wave, there are no low frequency cutoffs. The phase velocities of the flexural and screw modes decrease with increasing frequency. The fundamental modes have the high frequency tube wave velocity (or equivalently, the Scholte wave velocity, Schmitt, 1988b) as a high frequency limit; while the higher mode flexural and screw waves have the acoustic velocity of the borehole fluid as their high frequency asymptote. No matter whether we are exciting the fundamental or the higher order flexural or screw mode, the first arrival will always travel with the formation shear wave velocity, provided the source frequency band covers at least one of the modal cutoff frequencies. Thus, formation shear wave velocity can be picked by a straightforward, two-receiver moveout algorithm.

The excitation power of the dipole and quadrupole sources are lower than the monopole source (Schmitt, 1988b), especially with regards to the excitation of the low frequency tube wave and the corresponding fundamental flexural and screw modes. In general, successive higher order multipole sources (those with source pressure radiation pattern cosine $n\theta$, with $n$ an integer greater than 2) generate less and less energy in the respective modes.

Shear wave logging using flexural and screw modes has been successfully applied in the field. However, the nonsymmetric shear logging system is still not a routinely used and readily available tool. Nevertheless, it is the only tool which guarantees that one is measuring the true (or close to true) formation shear wave velocity in a slow formation. Other methods of measuring shear wave velocity in a slow formation all depend on indirect estimates and are model based. We describe some of these methods in the next section.

## 6.4    SHEAR VELOCITIES FROM GUIDED WAVES IN SLOW FORMATIONS

Slow formations are ones in which the shear wave velocity is lower than that of the acoustic wave velocity of the borehole fluid (typically drilling mud or water). In such formations, compressional or pressure wave energy generated in the borehole fluid column cannot be critically refracted along the borehole wall and radiate back into the borehole to be picked up by the receivers. Thus direct determination of shear wave velocity in a slow formation is not possible without the use of an nonaxisymmetric shear wave logging tool. Since most of the existing full waveform acoustic tools use the monopole source, it is necessary to derive indirect methods to determine formation shear wave velocity using the monopole full waveform data in slow formations. Before we start to discuss some of these methods, it is important to bear in mind that all these methods are interpretative methods based on specific models of the borehole. In general, all these methods are based on a cylindrical borehole in an elastic solid, with no irregularities on the borehole wall and no inhomogeneity in the formation. In actual applications, field conditions will be different from these idealized models. How different will determine the accuracy of the interpreted shear wave velocity. In particular, shales are not only characterized by slow seismic velocities, but they are also anisotropic. Interpretation methods applied to shales will at best give incomplete results and could potentially give totally erroneous information about the formation shear wave velocities. These considerations need to be kept in mind when applying the shear wave estimation methods given here.

### 6.4.1    Tube (Stoneley) Wave Interpretation

The first attempts at estimating formation shear wave velocity in a slow formation were made using the tube (Stoneley) wave (Cheng and Toksöz, 1982; Chen and Willen, 1984; Liu, 1984; Stevens and Day, 1986). The phase velocity of the tube wave, although not very sensitive to formation properties in a fast formation, is quite sensitive to the formation shear wave velocity in a slow formation. In such a formation, the high frequency tube wave velocity is mainly controlled by the formation shear wave velocity (Cheng et al., 1982; White, 1983; Chang et al., 1984). This influence of formation P wave velocity on the tube wave velocity is negligible. The other factor influencing the tube wave velocity is the borehole fluid. Thus, if we can measure the tube wave velocity and the acoustic velocity of the borehole fluid, we can solve for the formation shear wave velocity. If we do not know the acoustic velocity of the borehole fluid, then we must obtain the phase velocity of the tube wave at more than one frequency.

Using the measured velocity dispersion and the partition coefficients discussed later in Section 6.5.2, we can also solve for the tube wave velocity of the formation. This is basis of the inversion scheme used by Stevens and Day (1986).

## 6.4.2    Leaky P Mode Interpretation

Much of the full waveform data sets available were collected using the first generation full waveform acoustic logging tools. These tools typically have a narrow frequency band and a center frequency at around 13 to 15 kHz. At these frequencies, the tube wave is not very well excited in slow formations. Thus we cannot use the tube wave to estimate the formation shear wave velocity from data sets obtained with these narrow-band sources.

To address this problem, Cheng (1989) used the P leaky mode to invert for the shear wave velocity of a slow formation. This method is based on the dependence of P leaky mode amplitude on the Poisson's ratio. As the P head wave and P leaky mode propagate along the fluid-solid boundary of the borehole, energy is converted back into P waves in the fluid to be detected by the receiver, and is being radiated out into the formation as shear body waves in order to satisfy the boundary condition at the borehole wall. The amount of wave energy being converted into shear waves depends on the Poisson's ratio, and thus the shear wave velocity, of the information. Thus the P and P leaky mode amplitudes are dependent on both the formation P wave attenuation and the shear wave velocity. Cheng (1989) used the spectral amplitude ratio of the P wavetrain, which includes the P head wave and the P leaky modes, at two receivers to invert for the formation shear wave velocity and P wave attenuation. The forward model used in the inversion is the branch cut integral of Tsang and Rader (1979) which isolates the contribution of the P head wave and P leaky modes (Kurkjian, 1985). This method was tested on both synthetic and real data. It has been applied to a section of full waveform acoustic logs acquired in semiconsolidated sediments in DSDP (Deep Sea Drilling Project) hole 613. The results appear to be consistent with the formation shear wave velocities estimated using the differences in the P wave velocities calculated from the Wood's equation (Wood, 1930) assuming no shear contribution and those actually measured from the logs. This method appears to be unstable below a formation shear wave velocity of about 0.8 km/s.

## 6.5    ATTENUATION MEASUREMENTS IN BOREHOLES

One of the original promises of full waveform acoustic logging was the possibility of obtaining *in situ* measurements of formation attenuation ($1/Q$). Formation attenuation can be used for lithology and saturation interpretation. In general, a gas-saturated rock will show a higher $Q$ (low attenuation), and a water- or oil-saturated rock will show a lower $Q$ (higher attenuation). However, there is also evidence that partial gas saturation will result in a $Q$ lower than a fully fluid-saturated rock (Winkler and Nur, 1982). Fractured rocks, as well as rocks with high permeability, will have a low $Q$ (high attenuation).

*In situ* determination of attenuation, however, is much less straightforward than velocity determination. Any number of factors can influence the amplitude of the waveforms without affecting the observed arrival times of the various wave modes. Some of the more obvious ones are borehole irregularity, off-center tool, receiver mismatch, and geometric spreading associated with the head waves and leaky modes. In order to measure *in situ* attenuation properly, these effects have to be taken into account and corrected for. In this section, we will outline some of the techniques developed to measure *in situ* attenuation.

## 6.5.1    Attenuation Using Waveform Amplitudes

The most direct way of measuring formation P and S wave attenuation is the use of waveform amplitudes. This method assumes that we can isolate the P and S head waves. This is usually done by taking the first cycle of the respective wavetrains. With monopole excitation of regular full waveform logs and assuming far field radiation (more than a few wavelengths away), the P head wave has a geometric spending factor $1/z$, where $z$ is the source-receiver separation (Kurkjian, 1985; Zhang and Cheng, 1984). The S head wave has a geometric spreading factor of $1/z_2$ under the same conditions. Thus, if we assume that the receivers are matched, the tool is properly centralized, and the borehole is smooth, we can take the relative amplitudes of the P and S head waves at two receiver locations $z_1$ and $z_2$, correct for geometric spreading, and calculate the respective quality factors by the following formula:

$$\ln\frac{A(z_2)}{A(z_1)} = (z_2 - z_1)\frac{\pi f}{QV} \qquad\qquad 6.1$$

where $A$ is the amplitude of the wave, $f$ the frequency, and $V$ the velocity.

The problem with such an approach is that one cannot usually isolate the P and S head waves from the P leaky mode and pseudo-Rayleigh wave effectively, since the latter two modes travel at the P and S wave velocity at cutoff frequency. We are also using a very small portion of the full waveform. Finally, it is hard to estimate whether the receivers are in the far field, since the definition changes with frequency and wave velocity. More robust estimates of formation attenuation are made by formal inversion techniques utilizing the entire wavetrain. Two such methods are discussed next, one for shear wave attenuation and one for P wave attenuation.

## 6.5.2    Attenuation from Partition Coefficients

We first discuss the method of estimating *in situ* attenuation using partition coefficients. This method is used to determine formation shear wave attenuation using the pseudo-Rayleigh and tube waves (in case of a slow formation, just the tube wave). A detailed description is given in Burns and Cheng (1987). This method relies on the fact that the pseudo-Rayleigh and tube waves are normal modes of propagation in the borehole. It is thus possible to calculate analytically the way that wave energy is partitioned into acoustic vibrations in the fluid, and P and S particle motions in the formation. Using a perturbation method in combination with Hamilton's principle, one can calculate how the formation P and S wave velocities and the acoustic velocity of the borehole fluid affect the observed pseudo-Rayleigh and tube wave energy (Cheng et al., 1982). According to the correspondence principle, the pseudo-Rayleigh and tube wave attenuation are similarly affected (Aki and Richards, 1980). The pseudo-Rayleigh and tube wave attenuation are then given by a linear combination of the formation P and S wave attenuation and the borehole fluid attenuation. The coefficients in the linear combination are the normalized derivatives of the mode phase velocity $V_m$ (pseudo-Rayleigh and tube wave modes) with respect to the formation P and S wave velocity and the acoustic velocity of the borehole fluid; given in Equation 4.38:

$$\frac{1}{Q_m} = \frac{V_p}{V_m}\frac{\partial V_m}{\partial V_p}\frac{1}{Q_p} + \frac{V_s}{V_m}\frac{\partial V_m}{\partial V_s}\frac{1}{Q_s} + \frac{V_f}{V_m}\frac{\partial V_m}{\partial V_f}\frac{1}{Q_f} \qquad 4.38$$

This formula can be applied at either constant frequency or constant wavenumber, with the partial derivatives taken with the frequency or wavenumber taken constant. By using Hamilton's principle, these partial derivatives can be calculated analytically given the phase velocity, $c$. In general, the contribution of the formation P wave attenuation to the pseudo-Rayleigh and tube wave attenuation is negligible. Figures 6.9. and 6.10 show the partition coefficients for a sandstone formation and a soft sediment, respectively.

To determine pseudo-Rayleigh and tube wave attenuation, we first window the appropriate waveforms at two receiver locations, from the shear wave arrival time to the end of the record (Burns and Cheng, 1987). Figure 6.11 shows the windowing on synthetic microseismograms. This is more robust and stable than just taking the first cycle of the wavetrain. The windowed traces are then transformed into the frequency domain (Figure 6.12). The natural logarithm of the spectral amplitude ratio of the waveforms (Figure 6.13) can then be used in Equation 6.1 to obtain the attenuation of the wave as a function of frequency for the pseudo-Rayleigh and tube wave modes. These values can then be used in Equation 4.38 to solve for the S wave attenuation of the formation, together with the attenuation of the borehole fluid. Although in principle the formation P wave attenuation can also be obtained using this technique, the small values of the coefficient associated with it in Equation 4.38 make the solution very unstable and should not be used. Burns and Cheng (1987) applied this technique to synthetic data shown in Figure 6.11, with the inversion results in excellent agreement with the input model. They have also applied this technique to field

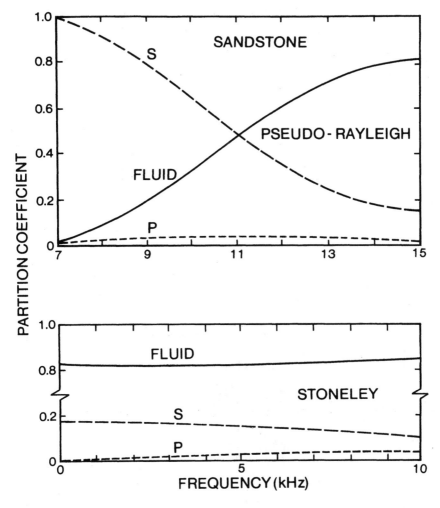

FIGURE 6.9.   Partition coefficients computed to evaluate the contributions of $Q_f$, $Q_p$, and $Q_s$ to attenuation of (A) first pseudo-Rayleigh mode, and (B) Stoneley mode; sandstone lithology with $V_p$ = 4.3 km/s and $V_s$ = 2.0 km/s.

data with qualified success, since no independent check on the formation attenuation was available.

As pointed out in an earlier section, the partition coefficients can also be used to invert for formation shear wave velocity from tube wave velocity dispersion. Rewriting Equation 4.38, we get:

$$\frac{\Delta V_m}{V_m} = \frac{V_p}{V_m}\frac{\partial V_m}{\partial V_p}\frac{\Delta V_p}{V_p} + \frac{V_s}{V_m}\frac{\partial V_m}{\partial V_s}\frac{\Delta V_s}{V_s} + \frac{V_f}{V_m}\frac{\partial V_m}{\partial V_f}\frac{\Delta V_f}{V_f} \qquad 6.2$$

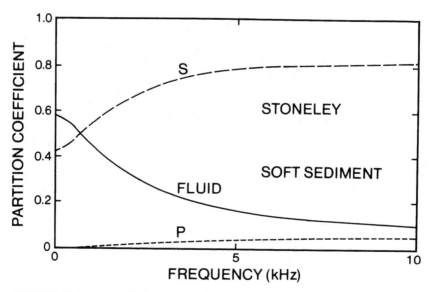

FIGURE 6.10. Partition coefficients computed to evaluate the contributions of $Q_f$, $Q_p$, and $Q_s$ to the attenuation of the Stoneley mode; soft sediment lithology with $V_p = 2.0$ km/s and $V_s = 0.8$ km/s.

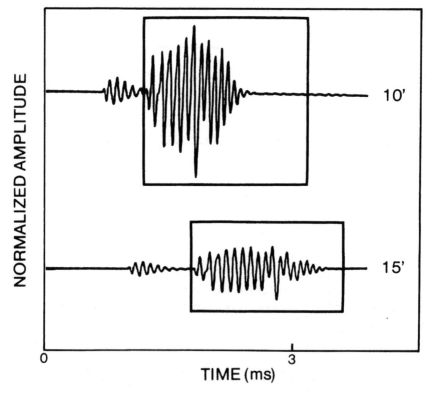

FIGURE 6.11. Synthetic microseismograms computed for sandstone lithology at source-to-receiver spacings of 3.0 and 4.5 m illustrating the windowing of the pseudo-Rayleigh mode in order to remove compressional wave energy from waveforms.

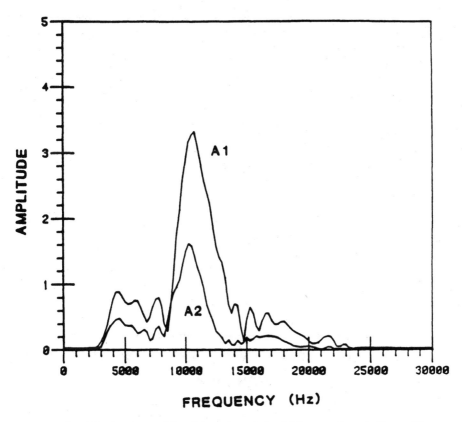

FIGURE 6.12. Transforms of the windowed pseudo-Rayleigh waves shown in Figure 6.11.

Equation 6.2 describes a linear relationship between the normalized changes in the phase velocity $V_m$ of the pseudo-Rayleigh and tube waves with the normalized changes in the formation and fluid velocities. This relationship can be used in a linearized iterative inversion algorithm to obtain the shear wave velocity from the phase velocity dispersion of the tube wave. Stevens and Day (1986) used this technique in their algorithm.

### 6.5.3    Attenuation Using Full Waveform Inversion of P Wavetrain

It is difficult to develop an analogous algorithm for formation P wave attenuation using the P leaky modes. The reason for this is that P leaky modes, unlike the pseudo-Rayleigh and tube wave, are not normal modes. It is thus more difficult and less straightforward to apply Hamilton's principle and develop the analytic expressions for partition coefficients in the P leaky mode case than in the normal case described in the previous section. Cheng (1989) instead used a full waveform inversion technique to obtain the formation P wave attenuation as well as S wave velocity. The method is based on the fact described earlier that the P head wave and P leaky mode are constantly leaky energy to converted S

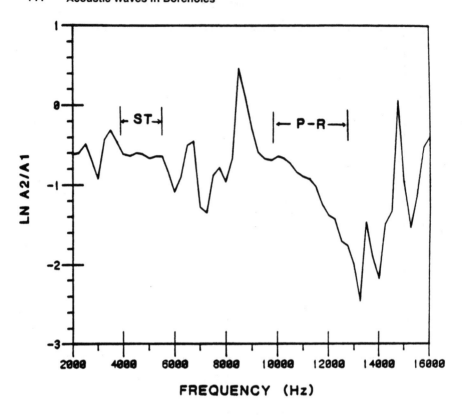

FIGURE 6.13. Spectral ratios given as a function of frequency for the spectra shown in figure 6.12.

wave radiating away from the borehole boundary. This leakage is modeled by the branch cut integral around the P wave branch point (Tsang and Rader, 1979; Kurkjian, 1985). The spectral ratio of the P head wave and P leaky mode packet at two receiver distances is used as data in a linearized iterative inversion scheme to estimate formation shear wave velocity and P wave attenuation. The branch cut integral is used as the forward model in the inversion. This method implicitly takes care of the geometrical spreading of the P wave packet by modeling it as leakage to shear wave propagating out into the formation. No far-field assumption has to be taken. As it turns out, this geometrical spreading is highly frequency dependent. If one takes the logarithm of the spectral ratio of the P wavetrain and applies it to Equation 6.1 to estimate formation P wave attenuation, assuming the geometrical spreading is constant in frequency, one gets a P wave attenuation that is too high because of the extra attenuation introduced by the nonconstant frequency geometrical spreading. This fact is demonstrated by the comparison of formation P wave attenuation obtained from a straight application of Equation 6.1 and laboratory measured P wave attenuation, with the latter much lower than the former (Goldberg and Zinszner, 1989). This can also be seen in the synthetic example shown in Figure 6.14.

a)

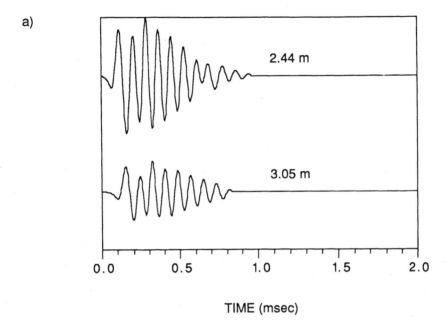

b)

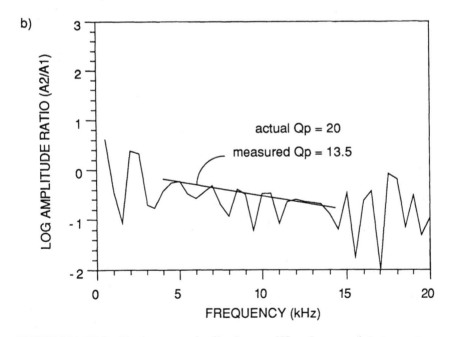

FIGURE 6.14. (A) Combination compressional head wave and PL mode wave packets at source-to-receiver spacings of 2.44 and 3.05 m computed for sandstone lithology; (B) spectral ratios for the two wave packets showing overestimation of attenuation (underestimation of $Q_p$) resulting from the spectral ratio method.

# 7

# Synthetic Seismograms and Acoustic Logging in Cased Boreholes

Acoustic logging is one of the logging methods that work in a cased hole environment, the other being nuclear logging. Historically, acoustic logging in cased boreholes had been used mainly as a casing evaluation tool. If the casing (pipe) is not properly bonded to the cement, a ringing signal will be generated by the acoustic log, with a transit time of 160 μsec/m, corresponding to the transit time of the plate mode of the steel casing. If the cement is bonded to the casing but not to the formation, the casing ringing will decrease, but may still be enough to obscure any formation arrival. When the casing is properly bonded to the cement and to the formation, P and S wave arrivals are usually identified. The introduction of the casing, however, changes the boundary condition between the solid and the liquid in the borehole. In this situation, the tube (Stoneley) wave is mainly controlled by the properties of the borehole fluid and the steel casing. Therefore, the effect of the formation on the tube wave velocity is negligible in cased holes, even in a soft formation.

In this chapter we will present the theory necessary to model seismic wave propagation in a cased borehole. These equations can also be used to model other changes in the borehole environment that are radially symmetric, such as an invaded zone or a damaged zone around the borehole. We will show examples of full waveform acoustic logs in cased boreholes. We will also present algorithms for determining formation velocities behind casing, especially in poorly cased boreholes. These algorithms will be illustrated using representative sets of full waveform data.

## 7.1 FORMULATING THE PROBLEM WITH CONCENTRIC LAYERS OF STEEL AND CEMENT

The problem of modeling acoustic logging in cased boreholes is addressed by several authors. The most comprehensive are the pair of papers by Tubman et al. (1984, 1986). We shall follow that development by expanding upon the theoretical formulation of elastic wave propagation in an open borehole given in Chapter 4. Much of that formulation is applicable in the cased hole problem. Consider a

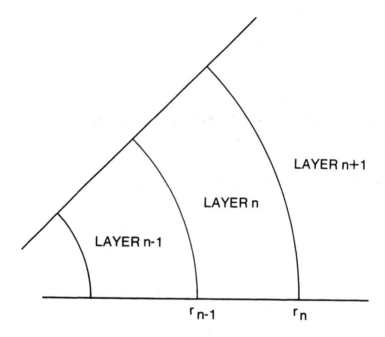

FIGURE 7.1.    Schematic illustration of multiple layers surrounding a fluid-filled borehole which
are used to model the fluid column, casing, cement, annulus, and formation.

borehole with multiple coaxial cylindrical layers shown in Figure 7.1. With loss
of generality we can assume an inner cylinder of fluid as the borehole. The case
with a logging tool can be taken into account in a procedure analogous to that
described in the first part of Chapter 4. The $n$th annulus has an inner radius of
$r_{n-1}$ and an outer radius of $r_n$. The outermost layer is infinite in extent and all layer
interfaces are considered welded contacts. From the definition of the displacement potentials in Equation 4.1, we can write the axial displacement in the $n$th
layer as

$$V_n = \frac{\partial \phi_n}{\partial z} + \frac{\partial (r\psi_n)}{r\partial r} \qquad 7.1$$

This, together with the radial displacement and stress and tangential stress given
by Equations 4.16 to 4.18, can be combined to form a stress-displacement vector
$u_n$ given by

$$u_n = \begin{pmatrix} u_n \\ v_n \\ \sigma_n \\ \tau_n \end{pmatrix} \qquad 7.2$$

$u$ and $v$ denote the radial and axial displacement and, where the radial and tangential stresses, $\sigma_{rr}$ and $\sigma_{rz}$, are represented by $\sigma$ and $\tau$, respectively. Substituting the forms for $\phi_n$ and $\psi_n$ from Equations 4.7 and 4.8 into Equations 4.16, 7.1, and 4.18 yields more explicit relationships between the displacements and stress and the potentials. By equating terms for each constant, the expressions can be combined into the form

$$u_n(r) = D_n(r)a_n \qquad 7.3$$

where

$$a_n = \begin{pmatrix} A_{1_n} \\ A_{2_n} \\ iB_{1_n} \\ iB_{2_n} \end{pmatrix} \qquad 7.4$$

and the elements of the $D$ matrix are given in Appendix 7.1.

We now have expressions for displacements and stresses for layer $n$ in terms of the potentials for that layer. A Thomsen-Haskell type propagator matrix (Thomsen, 1950; Haskell, 1953) is used to relate these across all the layers. The $n$th layer has an outer radius of $r_n$ and an inner radius of $r_{n-1}$, so, using Equation 7.3 at the outer radius, we have

$$u_n(r_n) = D_n(r_n)a_n \qquad 7.5a$$

and at the inner radius

$$u_n(r_{n-1}) = D_n(r_{n-1})a_n \qquad 7.5b$$

or

$$a_n = D_n^{-1}(r_{n-1})u_n(r_{n-1}) \qquad 7.5c$$

This displacement-stress vector $u_n(r_n)$ can then be related across the layer by combining Equations 7.5a and 7.5c and is stated as

$$u_n(r_n) = D_v(r_n)D_n^{-1}(r_{n-1})u_n(r_{n-1}) \qquad 7.6$$

This can be written as

$$u_n(r_n) = E_n(r_n, r_{n-1})u_n(r_{n-1}) \qquad 7.7$$

For the radially infinite, outermost layer $N$, Equation 7.5b gives

$$u_N\left(r_{N-1}\right) = D_N\left(r_{N-1}\right)a_N \qquad 7.8$$

Since displacements and stresses are continuous across the solid layer boundaries, we have

$$u_N\left(r_N\right) = E_{N-1}\left(r_{N-1}, r_{N-2}\right)E_{N-2}\left(r_{N-2}, r_{N-3}\right)...E_2\left(r_2, r_1\right)u_2\left(r_1\right) \qquad 7.9$$

or, by defining a matrix $G$, we have

$$Ga_N = u_2\left(r_1\right) \qquad 7.10$$

where

$$G = E_2^{-1}\left(r_2, r_1\right)...E_{N-1}^{-1}\left(r_{N-1}, r_{N-2}\right)D_N\left(r_{N-1}\right) \qquad 7.11$$

In the outermost layer, the radiation condition requires that there are no incoming waves, so $A_{2N} = B_{2N} = 0$, and we have

$$a_N = \begin{pmatrix} A_{1N} \\ 0 \\ iB_{1N} \\ 0 \end{pmatrix} \qquad 7.12$$

The requirements that the displacement and stress remain finite at $r = 0$ eliminates the $K_0$ and $K_0$ terms of the solutions in the central fluid cylinder ($A_{11} = B_{11} = 0$); and in a fluid, there is no vector potential, so $B_{21} = 0$. The displacement and stresses in the central fluid cylinder (at the fluid-solid boundary) can thus be written as

$$u_1\left(r_1\right) = \begin{pmatrix} u_1\left(r_1\right) \\ v_1\left(r_1\right) \\ \sigma_1\left(r_1\right) \\ \tau_1\left(r_1\right) \end{pmatrix} = \begin{pmatrix} D_{12}A_{21} \\ D_{22}A_{21} \\ D_{32}A_{21} \\ 0 \end{pmatrix} = \begin{pmatrix} m_1 I_1\left(m_1 r_1\right) \\ \kappa I_1\left(m_1 r_1\right) \\ -m_1\kappa^2 c^2 I_0\left(m_1 r_1\right) \\ 0 \end{pmatrix} A_{21} \qquad 7.13$$

At $r = r_1$, the fluid-solid interface, the radial displacement and radial stress are continuous, the axial displacement is discontinuous, and the tangential stress vanishes. Equating $u_2(r_1)$ to $u_1(r_1)$ and satisfying these boundary conditions by using the proper terms of Equations 7.12 and 7.13 and rearranging yields

$$\begin{pmatrix} m_1 I_1(m_1 r_1) & -G_{11} & -G_{13} \\ -m_1 \kappa^2 c^2 I_0(m_1 r_1) & -G_{31} & -G_{33} \\ 0 & -G_{41} & -G_{43} \end{pmatrix} \begin{pmatrix} A_{21} \\ A_{1N} \\ i B_{1N} \end{pmatrix} = 0 \qquad 7.14$$

The rows corresponding to the axial displacements are not included in the above equation because they are discontinuous and cannot be equated across the fluid-solid boundary. This is the period equation for waves traveling in the multilayered borehole. Dispersion relations are obtained by finding values of $\kappa$ and $c$ for which this relation holds true.

In the case of a solid-fluid or fluid-solid boundary between layers 2 to $N$, as in the case of a poorly bonded cased hole, the shear stress in the solid $\tau$ vanishes, and the matrix $\Delta$ can be solved explicitly to propagate the coefficients associated with the displacement potentials across the fluid layer. The details are given in Tubman et al. (1986) and in Appendix 7.2.

## 7.2    SYNTHETIC MICROSEISMOGRAMS FOR WELL BONDED AND POORLY BONDED EXAMPLES

To calculate the pressure response due to a point source in a multilayered borehole, we follow the development of Cheng et al. (1982) and Tubman et al. (1984), as well as Section 4.4 of this book. The pressure response $P_1(r,z,t)$ inside a borehole due to a point source $P_0(r,z,t)$ at $(0,0,0)$ is given by:

$$P(r,z,t) = \int_{-\infty}^{\infty} X(\omega) e^{-i\omega t} d\omega \int_{-\infty}^{\infty} A_{21} I_0(m_1 r) e^{i\kappa z} d\kappa \qquad 7.15$$

where $X(\omega)$ is the source spectrum. $A_{21}$ is the only nonzero constant associated with the potentials in the fluid layer. To determine $A_{21}$, a boundary condition is imposed at the borehole wall $r = r_1$. The specific condition is an expression for a $K_0(m_1 r)$ source in the frequency-wavenumber domain. This represents a point isotropic source on the borehole axis. This $K_0(m_1 r)$ source formulation is analogous to the $H_0^{(1)}(m_1 r)$ type of source used by Tsang and Rader (1979). The source term transforms Equation 7.14 into an inhomogeneous set of equations. Following the development above and in Section 4.4, and solving for $A_{21}$ yields

$$A_{21} = \frac{m_1 K_1 \left(m_1 (m_1 r_1)\right) F_1 - \rho_f \kappa^2 c^2 K_0 \left(m_1 r_1\right) F_2}{m_1 I_1 \left(m_1 r_1\right) F_1 + \rho_f \kappa^2 c^2 I_0 \left(m_1 r_1\right) F_2}$$    7.16

where

$$F_1 = G_{33} G_{41} - G_{31} G_{43},$$

and

$$F_2 = G_{13} G_{14} - G_{11} G_{34}.$$

The excitation resulting from the $K_0(m_1 r)$ source is added to the response function to give the total pressure field.

Equation 7.15 can be evaluated using the discrete wavenumber technique outlined in Section 4.4 of this book. Similarly, the effect of *in situ* attenuation for the P and S waves in the various layers can be taken into account by the introduction of complex velocities, also outlined in Section 4.4.

## 7.2.1    Well Bonded Case

The presence of the casing and cement can have a significant effect on the observed waveforms. Figure 7.2 shows synthetic microseismograms in a hard formation with and without casing. The character of the two microseismograms are entirely different. In particular, in the cased hole example, there is a strong tube wave which is absent in the open hole case. Both the P/P leaky wave and the S/pseudo-Rayleigh packet are smaller in amplitude in the cased hole. This can be attributed to the fact that the presence of the casing and cement effectively decrease the borehole radius, thus pushing the pseudo-Rayleigh and P leaky mode cutoff frequencies to a higher frequency. It is also apparent from the lower section of Figure 7.2 that both the P wave and the S wave arrive at the formation velocity, after correcting for the decreased travel time associated with the ray paths through the higher velocity cement and steel.

When the formation velocities are comparable to that of the steel, e.g., in a low porosity limestone, the resulting microseismogram shows a considerable amount of ringing. An example is shown in Figure 7.3. The steel and formation are fast layers on either side of the slow cement layer. It is interesting to note that, for the ringing portion of the time series, the thickness of the cement layer is one quarter of the dominant wavelength. The particular geometry and source frequency chosen here appear to cause some resonance effects in this layer.

In a slow formation, casing changes the character of the microseismogram even more drastically. Figure 7.4 shows the microseismograms in a cased and uncased borehole. The character in the cased borehole is similar to that in a fast formation, namely, it is dominated by the tube wave. Furthermore, the tube wave arrival time is very similar to that in a fast formation, reflecting the influence of

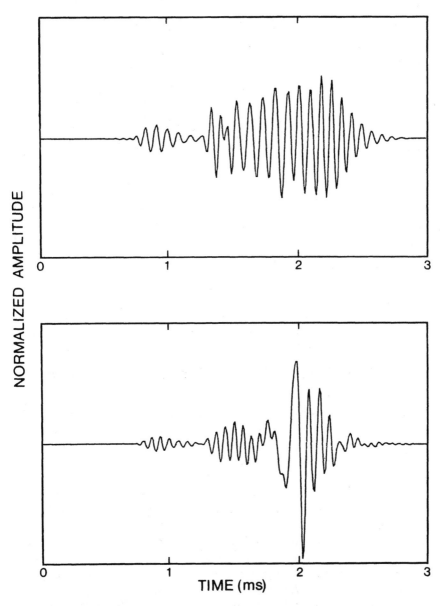

FIGURE 7.2.    Comparison of synthetic microseismograms computed for a sandstone lithology ($V_p$ = 4.88 km/s , $V_s$ = 2.60 km/s ; $R$ = 10 cm) (A) without casing and (B) with casing and cement included within the well bore ($R_c$ = 4.6 cm), and centerband frequency, $f_0$ = 13 kHz.

the steel casing instead of the formation itself. There is very little P wave energy and no S/pseudo-Rayleigh arrival. In contrast, in an open borehole, the P/P leaky wave packet is very strong, and the tube wave is prominent, but smaller in amplitude. The tube wave velocity is much slower than in the cased borehole and is influenced mainly by the formation S wave velocity.

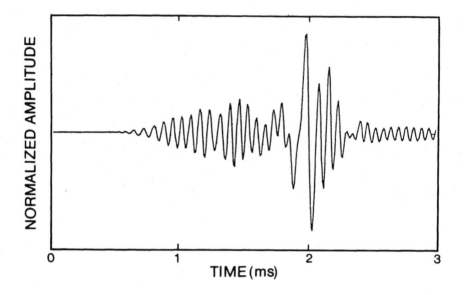

FIGURE 7.3. Synthetic microseismogram computed for low-porosity limestone lithology ($V_p$ = 5.93 km/s, $V_s$ = 3.20 km/s; $R$ = 10 cm, $R_c$ = 4.6 cm) where similarity in $V_p$ of rock and steel result in very ringy compressional arrivals; same source as Figure 7.2.

## 7.2.2    Poor Steel-Cement Bond, Good Cement-Formation Bond (Free Pipe)

In a poorly cemented cased hole, things get a little more complicated. The thickness of the fluid annulus plays an important role. We first look at the situation where there is no steel-cement bonding, but good cement-formation bond. This is known as the free pipe situation. This is modeled by inserting a layer of fluid between the solid steel casing and the cement.

Figure 7.5 shows the microseismograms for the free pipe situation for an array of receivers. The thickness of the fluid annulus is 1.27 cm. The formation is a fast formation, identical to the one used in Figure 7.2. There is a large, ringy P wavetrain, with a moveout of 5.5 km/s, or a transit time of 160 μsec/m, corresponding to the plate velocity of the steel. The formation S/pseudo-Rayleigh wave arrival is also not very coherent. The tube wave arrival, however, is still prominent. Figure 7.6 shows microseismograms of the free pipe situation in a formation with lower P and S wave velocity (but still a hard formation) typical of a porous sandstone. The casing arrival shows a little less but still substantial ringing, and the S/pseudo-Rayleigh packet is more coherent, but the formation velocities are still impossible to pick by hand, even with the help of an array of receivers.

The above examples were generated using a source center frequency of 13 kHz, typical of full waveform logging tools. Figure 7.7 shows an example of microseismograms generated with a source center frequency of 20 kHz to model a cement bond logging tool. In this case, all we can see is the high frequency

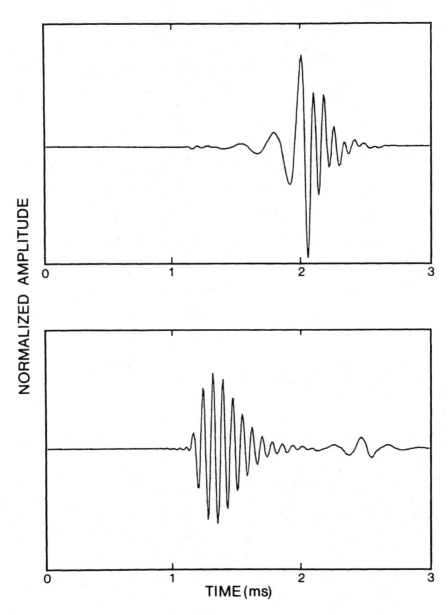

FIGURE 7.4.    Comparison of synthetic microseisograms computed for a shale lithology ($V_p$ = 2.90 km/s, $V_s$ = 1.52 km/s; $R$ = 10 cm) (A) without casing and (B) with casing and cement included within the well bore ($R_c$ = 4.6 cm); same source as Figure 7.2.

ringing of the casing arrival. At 20 kHz, which corresponds to a wavelength of 7.5 cm in water, the acoustic wave does not have enough penetration to excite any measurable formation arrival.

The effect of the thickness of the fluid annulus is shown in Figure 7.8. The thickness varies from 0 (no fluid layer, the well bonded case) to 4.45 cm in steps

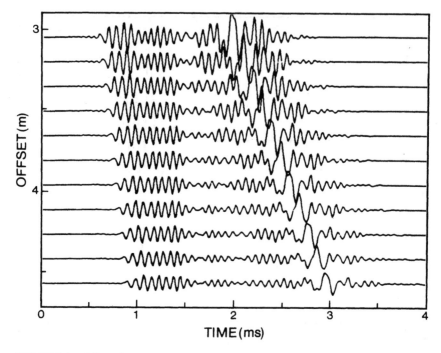

FIGURE 7.5.    Microseismograms computed for the sandstone formation in Figure 7.2 using the same source, except that there is a fluid annulus 1.27 cm thick between steel casing and cement ($R$ = 10 cm, $R_c$ = 4.6 cm).

of 0.64 cm. The cement thickness decreases correspondingly by an equal amount. The bottom microseismogram represents the case with a thick fluid annulus but no cement layer. The formation is the same as that in Figure 7.6. In all cases where there is a fluid layer, the casing arrival is prominent. The amplitude and duration of the ringing varies a little bit with the thickness of the fluid layer. In no case can the formation P wave arrival be readily identified. There is an extra tube wave introduced by the existence of the extra fluid annulus. However, the energy associated with it is very small (Tubman et al., 1986), and in general it is not evident on the computed microseismograms.

### 7.2.3    Good Steel-Cement Bond, Poor Cement-Formation Bond (Unbonded Casing)

The second class of poor bonding is when there is a good steel-cement bond but no cement-formation bond. We refer to this as the unbonded casing situation. The casing is now loaded with a coating of cement. This cement layer strongly affects the behavior of the pipe.

Figure 7.9 shows the unbonded casing situation with a thin layer of fluid (0.16 cm). The cement layer is 4.29 cm thick. The formation is the same as that in Figure 7.2. It is clear that the formation arrivals are easily identified. The cement has damped out the ringing of the casing. The pipe cannot generate a signal with

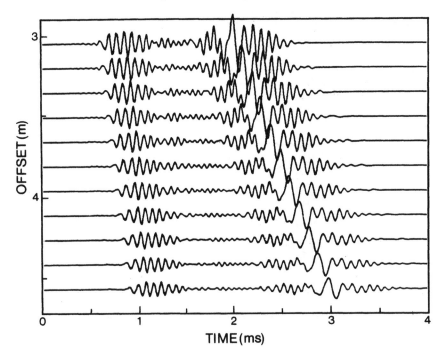

FIGURE 7.6.     Microseismograms computed for porous sandstone ($V_p$ = 4.0 km/s, $V_s$ = 2.13 km/s; $R$ = 10 cm, $R_c$ = 4.6 cm) for the same free pipe situation shown in Figure 7.5.

large amplitude and long duration as in the free pipe geometry. However, if we increase the thickness of the fluid annulus to 3.18 cm and decrease the thickness of the cement to 1.27 cm (Figure 7.10), the casing arrival once again becomes significant, although the character of the ringing changes substantially from the free pipe case (see Figure 7.5), reflecting the damping effect of the cement on the casing. The formation P wave arrival is now obscured, and the formation S/pseudo-Rayleigh wave arrival becomes less coherent and hard to identify.

The effect of the thickness of the fluid annulus and the cement layer is shown in Figure 7.11. The thickness of the fluid layer varies from 0 (no fluid layer, the well bonded case) to 4.45 cm in steps of 0.64 cm. The cement thickness decreases correspondingly by an equal amount. The bottom microseismogram represents the case with a thick fluid annulus but no cement layer. The formation is the same as that in Figure 7.6. This is analogous to Figure 7.8. As the thickness of the fluid layer increases and that of the cement decreases, the casing ringing becomes more obvious, and the formation P wave arrival becomes more obscure. At a fluid annulus thickness of 1.27 cm and a corresponding cement thickness of 3.18 cm, there is a hint of the casing arrival before the formation P wave arrival. As the thickness of the cement decreases further, the casing arrival becomes more prominent and earlier, reflecting the increasing effective velocity of the steel/cement combination. The tube wave arrival, however, is only slightly affected, demonstrating that in a cased hole the tube wave velocity is primarily controlled

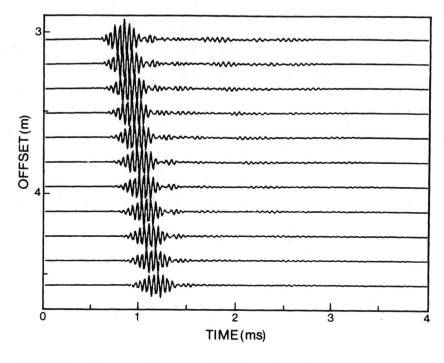

FIGURE 7.7.    Microseismograms computed for the same free pipe situation and the sandstone lithology in Figure 7.5, except that the source centerband frequency has been increased ($f_0 = 20$ kHz) to simulate the response of a conventional cement bond logging tool.

by the properties of the borehole fluid and the steel casing. Once again, the second tube wave, which exists in the fluid annulus, does not have enough energy to affect the observed microseismograms.

## 7.3  VELOCITY PICKING IN CASED HOLES — EXAMPLES USING SYNTHETIC MICROSEISMOGRAMS

As we have seen from the synthetic microseismogram examples in the previous section, there is usually coherent energy propagating at velocities characteristic formation properties in a cased hole, even a poorly cased one. The problem now is to measure formation P and S wave velocity from full waveform logs in a cased hole. Hsu and Baggeroer (1986) and Block et al. (1991) examined this problem using two different arrays processing velocity analysis techniques: the maximum likelihood method (MLM) and semblance cross-correlation (Kimball and Marzetta, 1984). In this section we will examine these two methods and test their accuracy using synthetic microseismograms generated by the discrete wavenumber summation method described in Section 7.2.

Both MLM and semblance velocity analysis are implemented within short time windows of given moveout across the receiver array. An example of such

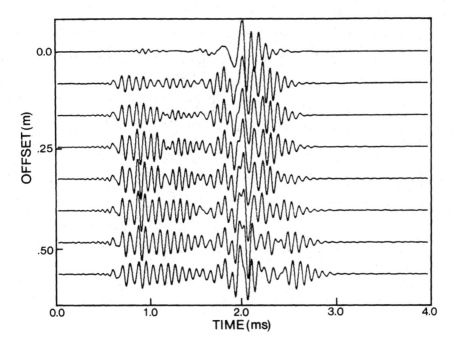

FIGURE 7.8.    Microseismograms computed using the lithology and borehole and casing diameters of Figure 7.6 to show the effects of fluid annulus thickness on wave propagation; the fluid annulus thickness increases from 0.0 (well bonded) at the top to 4.45 cm at the bottom, with the cement thickness decreasing accordingly.

a window is shown in Figure 7.12. $\tau$ is the beginning time of the window on the near trace, $T$ is the length of the window, and $n$ is the slope $(dx/dt)$ of the window. $n$ is equal to a trial phase velocity in the direction of the array and thus the phase velocity of propagation. For a fixed time $t$, calculations are made for a range of slownesses $p = dt/dx = 1/V$. The window is then advanced by a small amount $dt$, and the process is repeated. The final result is a series of contour plots as a function of time $\tau$ and velocity $V$ (or slowness $p$). The next two sections describe the two commonly used methods of velocity analysis: MLM and semblance.

### 7.3.1    Maximum Likelihood Method (MLM)

The maximum likelihood method essentially consists of performing a two dimensional Fourier transform in $(z,t)$ space for each slowness and starting time $(p,\tau)$ pair. The windowed traces are first Fourier transformed into the frequency domain to yield all the frequency components up to Nyquist. Let $x(t,z_\kappa)$ represent the time series at a distance $z_\kappa$. Then, at any given (or every) frequency $\omega_0$, a (spatial) spectral covariance matrix is formed, given by

$$K_x = \frac{XX^*}{T} \qquad\qquad 7.17$$

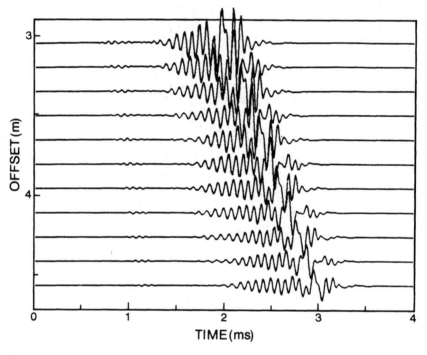

FIGURE 7.9.    Microseismograms computed for the sandstone lithology and borehole geometry of Figure 7.2, except that there is a 0.16 cm layer of fluid between the cement and formation to simulate the effects of poor bonding of cement to formation.

with

$$X = \begin{pmatrix} X\left(\tau, \omega_0, z_0\right) \\ X\left(\tau, \omega_0, z_1\right) \\ \cdot \\ \cdot \\ \cdot \\ X\left(\tau, \omega_0, z_{N-1}\right) \end{pmatrix}$$

7.18

$$X\left(\tau, \omega, z_\kappa\right) = e^{-i\omega z_\kappa p} \int_\tau^{\tau+T} x\left(t + z_\kappa p, z_\kappa\right) \, w(t-\tau) \, e^{-i\omega t} \, dt$$

7.19

and * denotes the complex conjugate transpose. $w(t)$ is the window function used in the time domain. In Block et al. (1986), this window function is given by

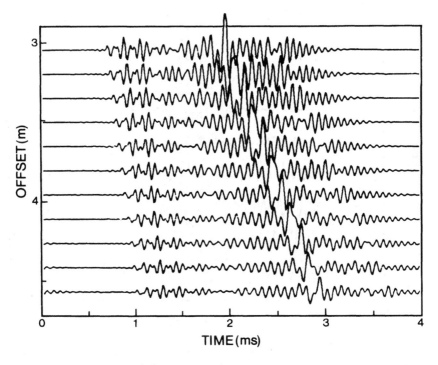

FIGURE 7.10.   Microseismograms computed for the same case as Figure 7.9, except that the fluid annulus has been increased to 3.18 cm.

$$w(t) = \sin^2\left(\frac{t\pi}{T}\right)$$   7.19a

The maximum likelihood estimate of the power due to the plane wave component traveling at frequency $\omega_0$ and slowness $p$ is then given by (Duckworth, 1983)

$$P(\tau,\omega_0,p) = \frac{1}{E * K_x^{-1} E}$$   7.20

with

$$E = \begin{pmatrix} e^{i\omega_0 p z_0} \\ e^{i\omega_0 p z_1} \\ . \\ . \\ . \\ e^{i\omega_0 p z_{N-1}} \end{pmatrix}$$   7.21

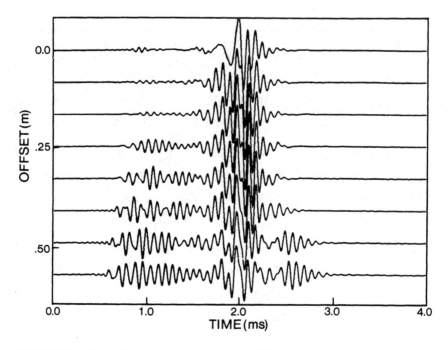

FIGURE 7.11. Microseismograms computed using the lithology and borehole and casing diameters of Figure 7.6 to show the effects of fluid annulus thickness when the annulus is located between the cement and formation; the fluid annulus thickness increases from 0.0 (well bonded) at the top to 4.45 cm at the bottom, with the cement thickness decreasing accordingly (compare to Figure 7.8).

The maximum likelihood estimate of the power is the one that provides an unbiased estimate of the power of the plane wave component at the scanning slowness $p$. In addition, the TOTAL power estimate within the time window is minimized in order to reduce the contributions from components with slownesses near $p$. For a complete discussion of this subject, the reader is referred to the original development given in Duckworth (1983). For the maximum likelihood method of velocity analysis, the output is power as a function of window starting time $\tau$ and slowness $p$ (or, more conveniently, velocity $n$) at discrete frequencies dictated by the time sampling $\Delta t$ and window length $T$. These frequencies correspond to $1/T$, $2/T$,...up to $1/2\Delta t$.

## 7.3.2  Semblance

The average semblance is essentially a correlation with amplitude taken into account. It is the ratio of the energy of a stacked trace to the sum of the energies of the individual traces within a time window, divided by the number of traces. Let $x(t,z_\kappa)$ represent the time series at distance $z_\kappa$. Then the average semblance $S$ within the time window $T$ at time $\tau$ with moveout $v = 1/p$ is given by

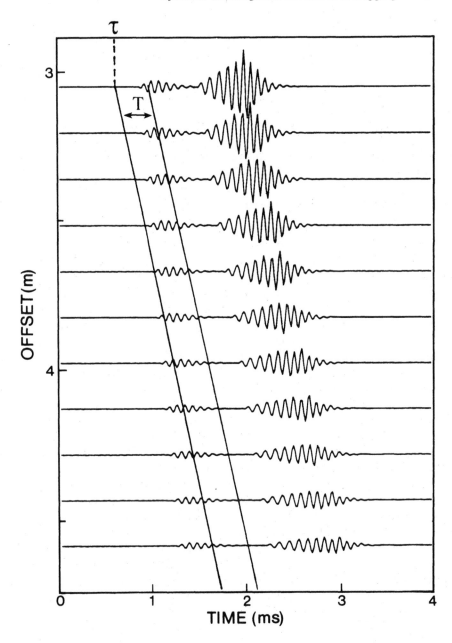

FIGURE 7.12.   Microseismograms computed to simulate wave mode arrivals across a receiver array, with time window and slope indicated to show how windows are configured for MLM and semblance velocity analysis.

$$S = \frac{\displaystyle\sum_{t_\kappa=\tau+pz_\kappa}^{\tau+pz_\kappa+T}\left(\sum_{\kappa=0}^{N-1}x\left(t_\kappa,z_\kappa\right)\right)^2}{N\displaystyle\sum_{t_\kappa=\tau+pz_\kappa}^{\tau+\pi\zeta_\kappa+T}\sum_{\kappa=0}^{N-1}\left(x\left(t_\kappa,z_\kappa\right)\right)^2}$$

7.22

The values of average semblance range from zero to one. These values are averaged over the entire frequency range up to the Nyquist frequency of $1/2\Delta t$ and are plotted as contours as a function of the window start time $\tau$ and moveout velocity $v$.

### 7.3.3    Open-Hole and Well-Bonded Case

We first test our velocity analysis algorithms with synthetic microseismograms generated in a simple open borehole. The formation velocities are $V_p = 4.0$ km/s, $V_s = 2.13$ km/s, and $V_f = 1.68$ km/s. Eleven traces were generated, with offsets ranging from 10 ft (3.05 m) to 15 ft (4.57 m) with a half foot (0.15 m) interval, and a sampling rate of 15.625 μsec. The microseismograms are shown in Figure 7.13. Figures 7.14A and 7.14B show the arrival time/moveout velocity plot for MLM and semblance, respectively. The MLM results were taken at a frequency of 4 kHz. Both methods show good results for the P and S arrivals, but the Stoneley wave arrival is not evident in the semblance plot.

The MLM depends quite a bit on the width of the window used. Figure 7.15 shows the results of using a 1 μsec window (crosses) and a 250 μsec window (squares). The line represents the theoretical dispersion curve for the pseudo-Rayleigh wave. As we can see, the velocities determined from a wider window follow the theoretical curve much better than that using a smaller window.

Both MLM and semblance methods give good results in the open borehole case. The accuracy and resolution of these methods depends on the size of the window and the digitization interval of the time traces. As in the velocity correlation schemes discussed in Chapter 6, interpolation of the waveforms or the semblance/MLM power estimates can be done to improve the velocity resolution of both techniques. The results from the well bonded cased hole are similar to the open hole case. The only exception is that when the formation P wave velocity is close to the steel velocity, the existence of the lower velocity cement layer results in more ringing in the microseismograms and an apparent P wave velocity analysis methods do a good job in determining formation velocities in open and well bonded based boreholes.

### 7.3.4    Free Pipe

In a free pipe situation, there is no steel-cement bond but there is good cement-formation bond. In this case, the ringing from the casing tends to mask the formation arrivals. Figure 7.16 shows the microseismograms of a free pipe model with the same formation velocities as Figure 7.14. The formation P wave arrival is not evident. Figures 7.17A and 7.17B show the results of the MLM at

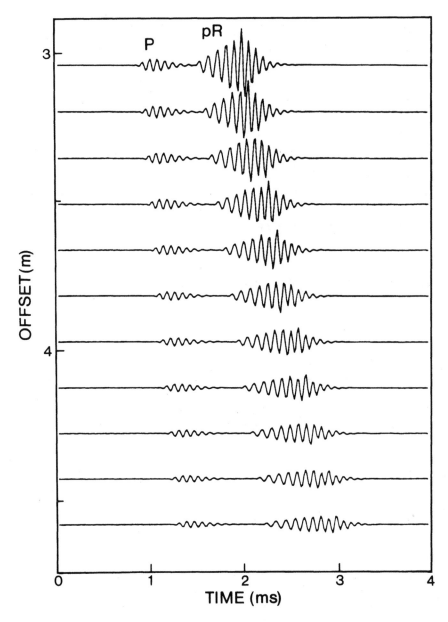

FIGURE 7.13.    Microseismograms computed for an array of receivers in a sandstone formation ($V_p$ = 4.0 km/s, $V_s$ = 2.13 km/s, $V_f$ = 1.68 km/s) and well-bonded conditions ($R$ = 10 cm, $R_c$ = 4.6 cm).

12 kHz and semblance, respectively. Both results are similar, with slightly lowered estimates of formation velocities (especially for the P wave) than actual. Semblance gives a better estimate of the steel velocity at 5.35 km/s. The differences in the estimated and actual velocities (for both the P and the S) for both methods are about 17 µs/m. Given a digitization interval of 15.625 µs, an

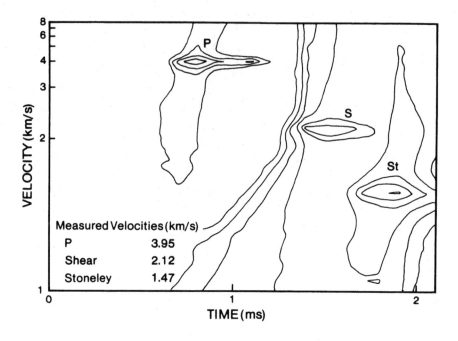

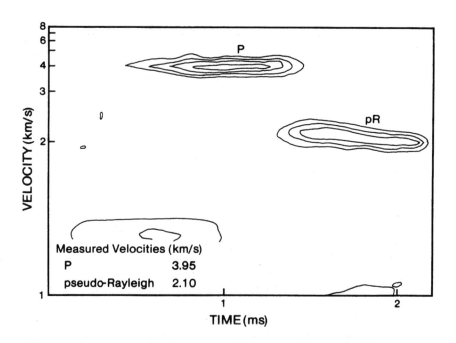

FIGURE 7.14. Arrival time/moveout velocity plot for (A) MLM method at 4 kHz and (B) semblance method computed for the waveforms in Figure 7.13.

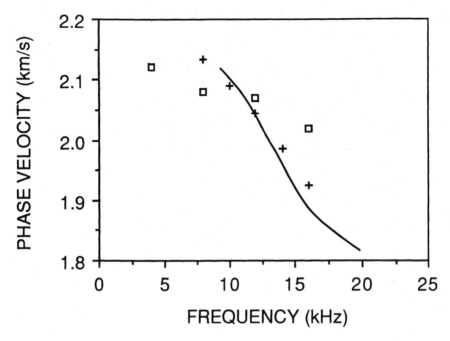

FIGURE 7.15. Dispersion curves for the pseudo-Rayleigh wave computed using the waveform data in Figure 7.13 using the MLM method with 1.0 and 0.250 ms time windows illustrating the effect of time window on velocity picks.

array length of 1.68 m (5.5 ft), and no interpolation of the final results, this magnitude of error can be expected. For commercially available tools, the sampling interval is usually 5 μs, and should improve the accuracy of the formation velocities determined behind free pipe.

### 7.3.5    Unbonded Casing

The unbonded casing is one where the cement is bonded to the steel casing but not to the formation. As discussed earlier in this chapter, the ringing characteristics of the combined steel-cement mixture vary, depending on the thickness of the cement. Figure 7.18 shows the synthetic microseismogram for a unbounded cased hole with formation velocities equal to those in Figure 7.14. Figures 7.19A and 7.19B show the MLM results at 11.7 kHz and semblance results, respectively. In both cases, the formation P wave cannot be separated from the combined steel-cement casing arrival. The formation S wave velocity is quite well determined. Numerical experiments have shown that the formation P wave velocity above about 30 km/s cannot be resolved easily in an unbonded cased hole. Block et al. (1991) performed several numerical experiments to determine the ability of MLM and semblance to determine formation velocities behind casing in poorly bonded cased holes. Their results are summarized here in Table 7.1.

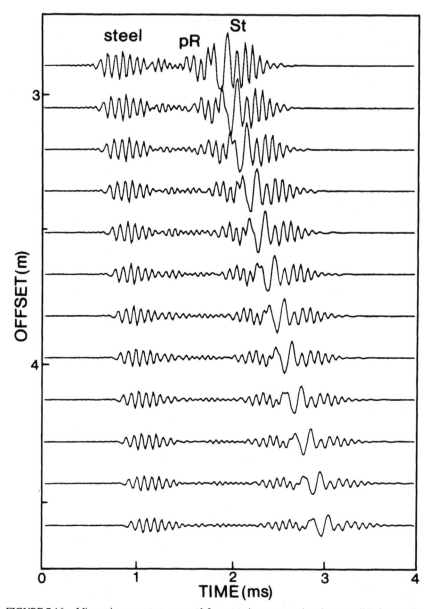

FIGURE 7.16.   Microseismograms computed for a receiver array using the same lithology and borehole conditions as in Figure 7.13 except that the free pipe situation has been modeled using a 1.27 cm thick annulus between the casing and cement.

## 7.4   VELOCITY PICKING IN CASING HOLES — FIELD EXAMPLES

How well do these two velocity analysis techniques work under actual field

**Table 7.1**
**Summary of the Major Conclusions from Analysis of Synthetic Acoustic Logging Data for Free-Pipe and Unbonded Casing Situations**

| Formation | Free pipe | | Unbonded casing | |
|---|---|---|---|---|
| Velocities | MLM | Semblance | MLM | Semblance |
| Very slow (no S wave) | OK[a] | OK[a] | OK | OK |
| Slow | OK[a] | OK[a] | Cannot separate P wave from casing arrival | Weak but distinguishable P wave maximum |
| Moderate | OK | OK | Cannot separate P wave from casing arrival | Cannot separate P wave from casing arrival |
| Fast | OK | Filtering required to separate P wave from casing arrival | Cannot separate P wave from casing arrival | Cannot separate P wave from casing arrival |
| Very fast (P wave velocity $\approx$ 6 km/s) | Cannot separate P wave from casing arrival | Cannot separate P wave from casing arrival | Cannot separate P wave from casing arrival | Cannot separate P wave from casing arrival |

[a] Situations that were not modeled — conclusions were deduced from other results.

conditions? Hsu and Baggeroer (1986) showed results from an experimental 12 receiver tool. In this section, we will show the results presented by Block et al. (1991) from a production 8 receiver tool (Schlumberger SDT™) and data collected at the test borehole of the Reservoir Delineation Group of the Earth Resources Laboratory at M.I.T. The borehole is located just Southwest of Traverse City, in the Michigan Reef trend. We will present the results from both the well bonded and the poorly cased hole situation.

## 7.4.1    Well Bonded Case

Figure 7.20 shows the open hole waveforms from a low porosity limestone section. There appears to be quite a bit of variation in the amplitudes across the receivers which could be due to poor receiver equalization. Figures 7.21A and 7.21B show the results from MLM at 7.8 kHz and semblance analysis, respectively. The P wave velocity of 6.2 km/s is well defined, as is the S wave velocity of 3.2 km/s. In the MLM plot there is energy at 2.8 km/s that could be related to the Airy phase of the pseudo-Rayleigh wave.

Figure 7.22 shows the waveforms from the cased hole at the same depth after the casing was put in. This section of the hole is considered to be well cemented, as can be seen from the lack of ringing in the P wave waveform. Figures 7.23A and 7.23B show the results from MLM at 7.8 kHz and semblance analysis, respectively. The resolution of the MLM is degraded compared to the open hole case, and the S/pseudo-Rayleigh arrival peak is broader and harder to pick a single velocity from. The semblance results identify the P wave very well, but the S/pseudo-Rayleigh suffers from the same lack of resolution as in the MLM case. Nevertheless, both P and S wave velocities obtained are consistent with those from the open hole waveforms.

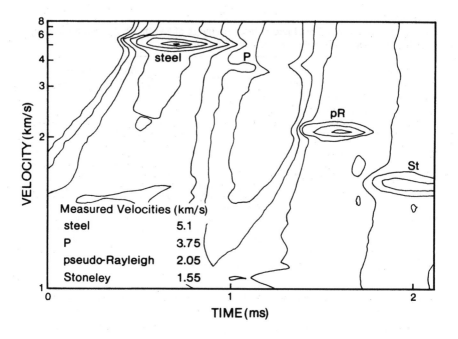

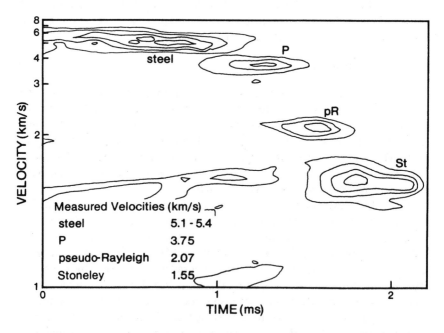

FIGURE 7.17.   Arrival time/moveout velocity plot for (A) MLM method at 4 kHz, and (B) semblance method computed for the waveforms in Figure 7.16 (compare to Figure 7.14).

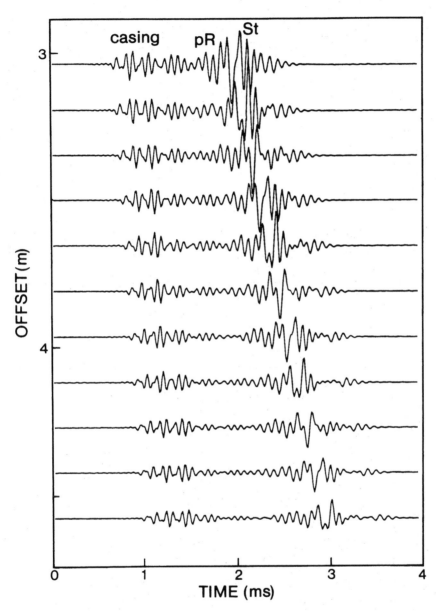

FIGURE 7.18. Microseismograms computed for a receiver array using the same lithology and borehole conditions as Figure 7.13 except that the free casing situation has been modeled using a 3.18 cm thick fluid annulus between the cement and formation.

## 7.4.2    Poor Bonded Case

Figure 7.24 shows the open hole waveforms from a low porosity sandstone section of the M.I.T. borehole in Michigan. Figure 7.25A and 7.25B show the

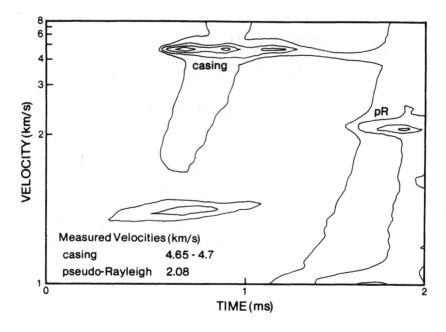

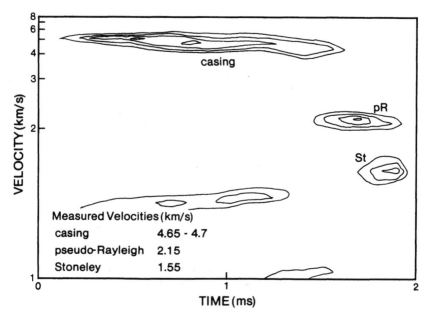

FIGURE 7.19.  Arrival time/moveout velocity plot for (A) MLM method at 4 kHz and (B) semblance method computed for the waveforms in Figure 7.18 (compare to Figures 7.14 and 7.17).

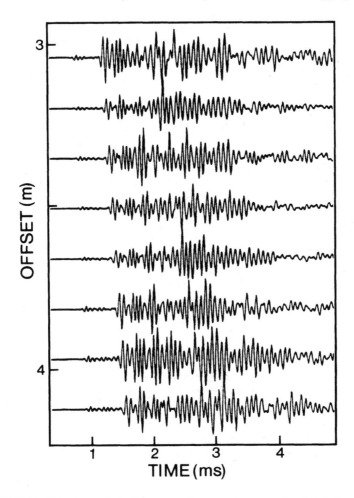

FIGURE 7.20. Waveforms obtained from a receiver array in an open borehole in low-porosity limestone; borehole fluid is water and nominal hole diameter is 10 cm.

results from MLM at 7.8 kHz and semblance, respectively. The MLM results show a P wave velocity of about 5.6 km/s and a S wave velocity of about 3 km/s, with perhaps an indication of an Airy phase. The Stoneley wave is also quite evident at about 1.45 km/s. The semblance results show a slightly lower P wave velocity of 5.5 km/s. The S and Stoneley wave velocities are similar to those determined by the MLM.

Figure 7.26 shows the waveforms at the same interval after casing. No cement was used in this interval. The waveforms show characteristic ringing of the free pipe. Figures 7.27A and 7.27B show the results from MLM at 7.8 kHz and semblance, respectively. There are two distinct arrivals around the P wave velocity of 5.5 km/s, one slightly slower at 5.0 km/s. The origin of this lower peak

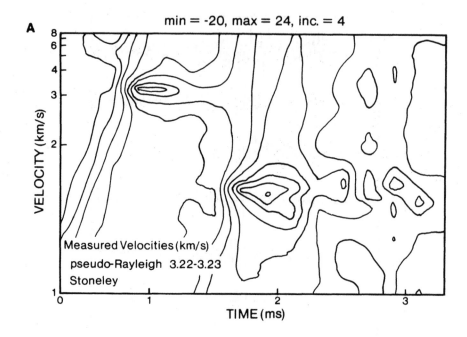

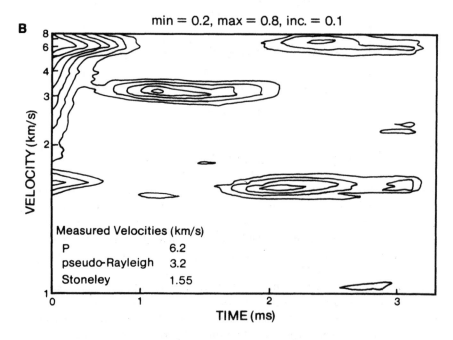

FIGURE 7.21.   Arrival time/moveout velocity plots for (A) MLM method at 7.8 kHz, and (B) semblance method applied to the waveform data in Figure 7.20.

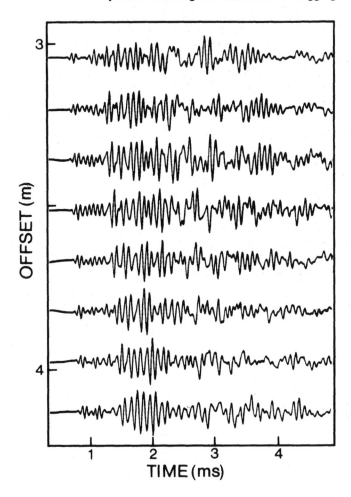

FIGURE 7.22.   Waveforms obtained from a receiver array in the same borehole as in Figure 7.20 after installation of casing ($R_c$ = 8.4 cm).

is not well understood. Because the MLM results are for a frequency of 7.8 kHz, the casing ringing does not affect them too much. The S wave velocity can be identified at 2.9 to 2.95 km/s, and the Stoneley wave velocity at 1.53 km/s. In contrast, for the semblance, because it is a time domain method and takes into account all frequencies, the casing ringing stands out. The ringing also degrades the results for the S wave and Stoneley wave velocities. However, if we apply a low pass filter to the data prior to performing the semblance analysis, the results are much better. Figure 7.28A shows the results of filtering the waveforms using a low pass filter with a cutoff of 10 kHz and Figure 7.28B the results of performing the semblance analysis. The casing ringing has disappeared, and P wave energy is easily identified from the semblance results. The S and Stoneley wave velocities are also easily determined.

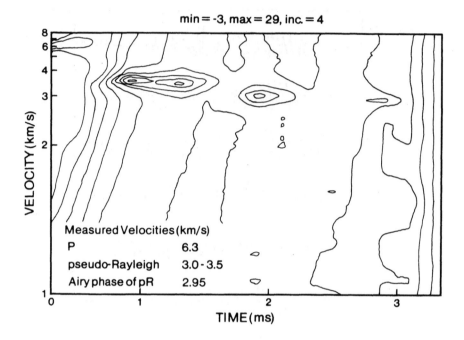

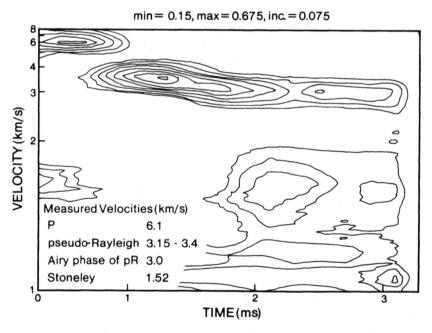

FIGURE 7.23.   Arrival time/moveout velocity plots for (A) MLM method at 7.8 kHz and (B) semblance method applied to the wavefom data in Figure 7.22.

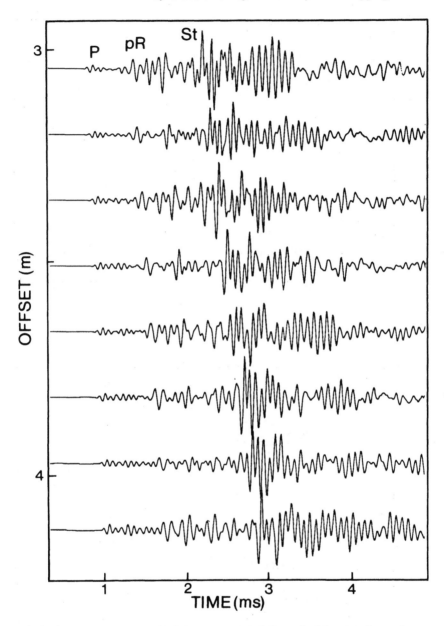

FIGURE 7.24. Waveforms obtained from an open hole interval of low-porosity sandstone; borehole fluid is water and nominal borehole diameter is 10 cm.

## 7.5    ATTENUATION MEASUREMENTS IN CASED BOREHOLES

Attenuation measurements in cased boreholes are very similar to the procedures used for the open borehole. The exception is in the formation P wave

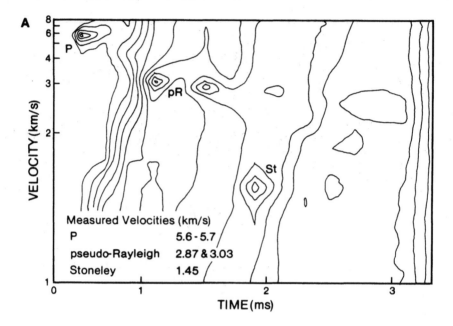

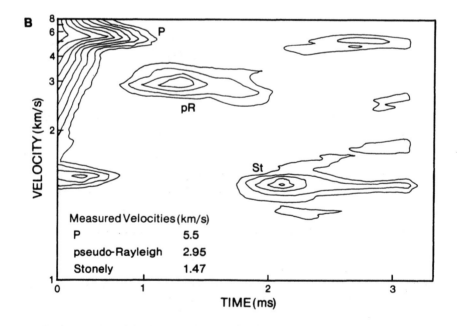

FIGURE 7.25.   Arrival time/moveout velocity plots for (A) MLM method at 7.8 kHz and (B) semblance method applied to the waveform data in Figure 7.24.

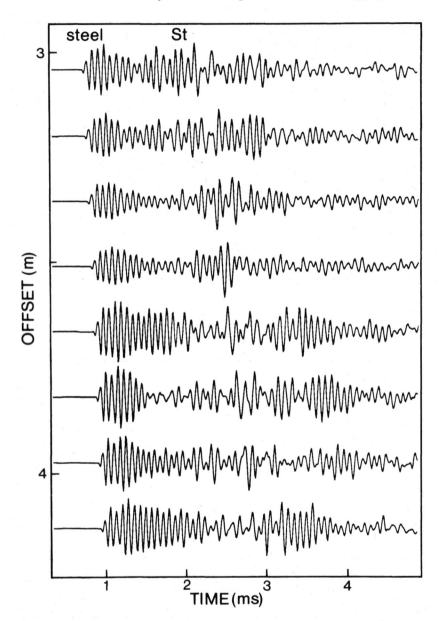

FIGURE 7.26.   Waveforms obtained in the same interval as Figure 7.24 after intsallation of casing $(R_c = 8.4 \text{ cm})$.

attenuation measurements. Because of the interference of the casing modes in the free pipe and unbounded casing situations, no effective methods of determining formation P and S wave attenuation behind casing have been developed to date. For the well bonded case, the formalism for estimating formation P wave attenuation has not been developed, although in principle an algorithm based on

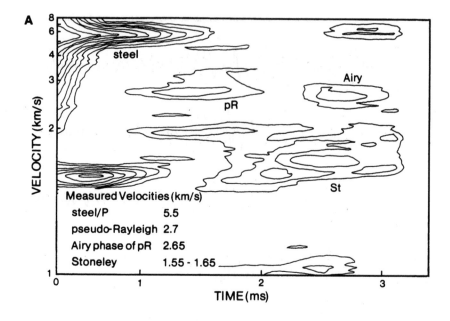

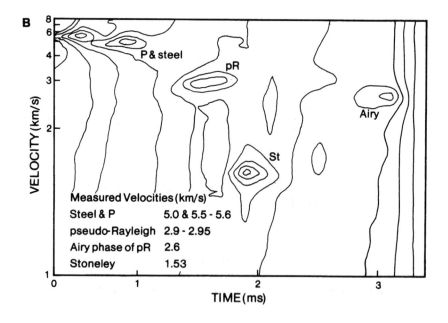

FIGURE 7.27.   Arrival time/moveout velocity plots for (A) MLM method at 7.8 kHz and (B) semblance method applied to the waveform data in Figure 7.26.

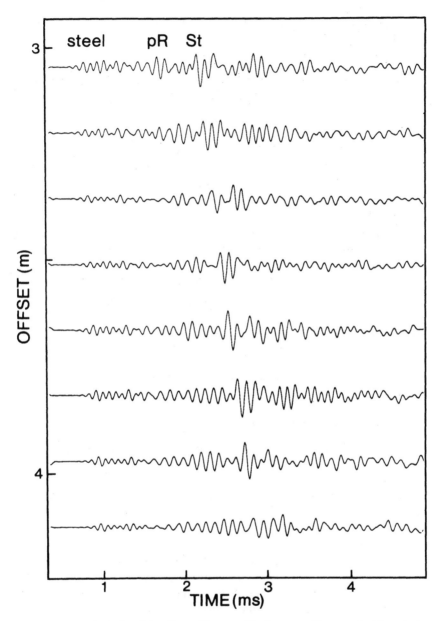

FIGURE 7.28a. Illustration of the effects of low-pass filtering on semblance method for velocity picking: waveforms in Figure 7.26 after application of a 10 kHz low-pass filter.

the full waveform inversion using the branch cut integration technique for the open hole (see Section 6.5.3) can be adopted to the well bonded cased hole. For formation S wave attenuation, Burns and Cheng (1987) utilized the partition coefficient formulation to estimate formation S wave attenuation from the

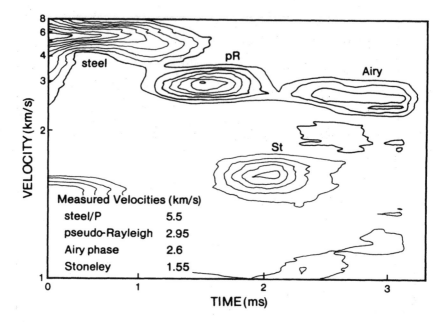

FIGURE 7.28b. Illustration of the effects of low-pass filtering on semblance method for velocity picking: arrival time/moveout velocity plots for semblance method applied to filtered waveforms.

pseudo-Rayleigh and Stoneley waves. The Stoneley wave is mainly sensitive to the casing and borehole fluid parameters.

### 7.5.1    Partition Coefficients

The principles behind the use of partition coefficients have been described previously in Sections 4.6 and 6.5.2 and will not be repeated here. The major difference is that now we have to consider the attenuation of the cement and steel casing in addition to the formation and borehole fluid attenuation. As in the open hole case, formation P wave attenuation, cement P wave attenuation, and steel P wave attenuation have negligible contributions to the total pseudo-Rayleigh and Stoneley wave attenuation.

Figure 7.29 shows the partition coefficients for the pseudo-Rayleigh wave for a well cased borehole. At around the cutoff frequency of 13 kHz, the pseudo-Rayleigh wave is most sensitive to the formation S wave. However, the sensitivity drops off quickly as the frequency increases. Above 18 kHz the pseudo-Rayleigh wave is sensitive to the borehole fluid and the S waves of the casing and cement. This sharp decrease in the partition coefficient for the information S wave makes the proper determination of the cutoff frequency critical in the estimation of the formation S wave attenuation.

### 7.5.2    Synthetic Example

Figures 7.30A, 7.30B, and 7.30C show the synthetic microseismograms,

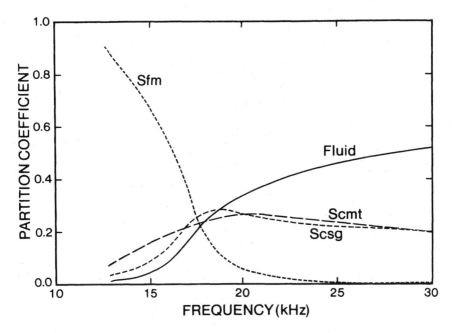

FIGURE 7.29.  Partition coefficients computed for a cased borehole in a sandstone formation ($V_p$ = 4.88 km/s, $V_s$ = 2.60 km/s, $R$ = 10 cm, $R_c$ = 4.6 cm; borehole fluid is water).

amplitude spectra and spectral ratios from a well bonded cased hole used in the inversion for formation S wave attenuation. The start of the window used is also shown in Figure 7.30A. The frequency ranges used for the Stoneley and pseudo-Rayleigh wave in the inversion are shown in Figure 7.30C. The inversion results are good for the borehole fluid and cement S wave attenuation. However, the formation S wave attenuation was off by 25%. It appears that the former two are very well constrained by the Stoneley wave attenuation. The formation S wave attenuation, however, is critically dependent on the location of the cutoff frequency of the pseudo-Rayleigh wave. In this example, the theoretical cutoff frequency is about 13 kHz (see Figure 7.29). However, from the spectral ratio plot (Figure 7.30C), the cutoff frequency from the synthetic microseismogram appears to be around 15 kHz. On examination of the actual spectral amplitudes, it is apparent that the peak of the pseudo-Rayleigh wave energy is around 12 to 13 kHz, but there are notches in the spectra. Thus a straight spectral ratio calculation results in widely oscillatory behavior from 12 to 15 kHz, and in particular, producing a sharp peak at 15 kHz corresponding to the notch in the spectrum of the near trace A1. A more robust estimator of the spectral ratio such as a Weiner deconvolution algorithm may help to stabilize these fluctuations and improve the inversion results.

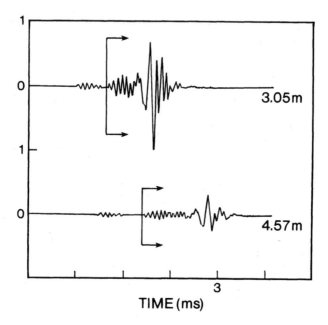

FIGURE 7.30.   Inversion for shear attenuation $(1/Q_s)$ illustrated using synthetic microseismogram data for the same cased borehole given in Figure 7.29: (A) microseismograms for 3.0 and 4.6 m offsets showing start of time window used, (B) amplitude spectra obtained from the windowed waveforms, and (C) amplitude ratios for the two waveforms showing portions of spectra applicable to Stonely and pseudo-Rayleigh modes; $Q_s$ is interpreted from these ratios and the calculated partition coefficients.

# Appendix 7.1

## Elements of the D Matrix:

$$D_{n11}(r) = -m_{2n}K_1(m_{2n}r)$$

$$D_{n12}(r) = m_{2n}I_1(m_{2n}r)$$

$$D_{n13}(r) = -\kappa K_1(m_{3n}r)$$

$$D_{n14}(r) = -\kappa I_1(m_{3n}r)$$

$$D_{n21}(r) = \kappa K_0(m_{2n}r)$$

$$D_{n22}(r) = \kappa I_0(m_{2n}r)$$

$$D_{n23}(r) = m_{3n}K_0(m_{3n}r)$$

$$D_{n24}(r) = -m_{3n}I_0(m_{3n}r)$$

$$D_{n31}(r) = \rho_n\kappa^2(V_s^2-c^2)K_0(m_{2n}r) + 2\rho_nV_s^2\frac{m_{2n}^2}{r}K_1(m_{2n}r)$$

$$D_{n32}(r) = \rho_n\kappa^2(V_s^2-c^2)I_0(m_{2n}r) - 2\rho_nV_s^2\frac{m_{2n}^2}{r}I_1(m_{2n}r)$$

$$D_{n33}(r) = -2\rho_nV_s^2\kappa m_{3n}\left[K_0(m_{3n}r) + \frac{1}{m_{3n}r}K_1(m_{3n}r)\right]$$

$$D_{n34}(r) = 2\rho_nV_s^{2\kappa m_{3n}}\left[I_0(m_{3n}r) - \frac{1}{m_{3n}r}I_1(m_{3n}r)\right]$$

$$D_{n41}(r) = -2\rho_nV_\sigma^2\kappa m_{2n}K_1(m_{2n}r)$$

$$D_{n42}(r) = 2\rho_nV_s^2\kappa m_{2n}I_1(m_{2n}r)$$

$$D_{n43}(r) = -\rho_n\kappa^2(V_s^2-c^2)K_1(m_{3n}r)$$

$$D_{n44}(r) = -\rho_n\kappa^2(V_s^2-c^2)I_1(m_{3n}r)$$

A7.1.1

As pointed out by Schmitt and Bouchon (1985), there were three typographical errors in the original formulas given by Tubman et al. (1984). Schmitt and

Bouchon also pointed out that since the determinant of $D_n$ is just given (after applying the Wronskian of the modified Bessel functions) by

$$Det\{D_n(r)\} = -\frac{\omega^4 \rho_n^2}{r^2}$$

*A7.1.2*

So, by applying Cramer's rule, the inverse matrix is just given by

$$D_n^{-1}(r) = \frac{r}{\omega^2 \rho_n} \times \begin{pmatrix} D_{n32}(r) & D_{n42}(r) & -D_{n12}(r) & -D_{n22}(r) \\ -D_{n31}(r) & -D_{n41}(r) & D_{n11}(r) & D_{n21}(r) \\ D_{n34}(r) & D_{n44}(r) & -D_{n14}(r) & -D_{n24}(r) \\ -D_{n33}(r) & -D_{n43}(r) & D_{n13}(r) & D_{n23}(r) \end{pmatrix}$$

*A7.1.3*

# Appendix 7.2

## Boundary Conditions for Poorly Bonded Cased Holes

### Fluid-Solid Boundary

Let layer $n$ be the inner fluid layer and $n+1$ be the outer solid layer. The boundary is then located at $r = r_n$. The boundary conditions are the continuity of radial displacement and stress and zero axial stress at the boundary wall. The axial displacements are discontinuous. The boundary conditions can be written as

$$u_n(r_n) = u_{n+1}(r_n),$$
$$\sigma_n(r_n) = \sigma_{n+1}(r_n),$$
$$\text{and} \qquad \tau_n(r_n) = \tau_{n+1}(r_n), = 0 \qquad\qquad A7.2.1$$

In the fluid layer, the constants to be solved for are $A_{1n}$ and $A_{2n}$, while in the solid layer, we have four constants: $A_{1n+1}$, $A_{2n+1}$, $B_{1n+1}$, and $B_{2n+1}$. In the solid, Equation 7.3 becomes, at $r = r_n$,

$$\begin{pmatrix} u_{n+1} \\ v_{n+1} \\ \sigma_{n+1} \\ 0 \end{pmatrix} = \begin{pmatrix} D_{n+1_{11}} & D_{n+1_{12}} & D_{n+1_{13}} & D_{n+1_{14}} \\ D_{n+1_{21}} & D_{n+1_{22}} & D_{n+1_{23}} & D_{n+1_{24}} \\ D_{n+1_{31}} & D_{n+1_{32}} & D_{n+1_{33}} & D_{n+1_{34}} \\ D_{n+1_{41}} & D_{n+1_{42}} & D_{n+1_{43}} & D_{n+1_{44}} \end{pmatrix} \begin{pmatrix} A_{1n+1} \\ A_{2n+1} \\ iB_{1n+1} \\ iB_{2n+1} \end{pmatrix} \qquad A7.2.2$$

In the fluid, at $r = r_n$, we have

$$\begin{pmatrix} A_{1n} \\ A_{2n} \end{pmatrix} = \begin{pmatrix} D_{n_{11}}^{-1} & D_{n_{13}}^{-1} \\ D_{n_{21}}^{-1} & D_{n_{23}}^{-1} \end{pmatrix} \begin{pmatrix} u_n \\ \sigma_n \end{pmatrix} \qquad A7.2.3$$

Using the first and third subequation in Equation A7.2.2, namely,

$$\begin{pmatrix} u_{n+1} \\ \sigma_{n+1} \end{pmatrix} = \begin{pmatrix} D_{n+1_{11}} & D_{n+1_{12}} & D_{n+1_{13}} & D_{n+1_{14}} \\ D_{n+1_{31}} & D_{n+1_{32}} & D_{n+1_{33}} & D_{n+1_{34}} \end{pmatrix} \begin{pmatrix} A_{1n+1} \\ A_{2n+1} \\ iB_{1n+1} \\ iB_{2n+1} \end{pmatrix} \qquad A7.2.4$$

and combining it with Equation A7.2.3, making use of Equation A7.2.1, we can express the column vector $a_n$, which contain only two unknowns, $A_{1n}$ and $A_{2n}$, in terms of $a_{n+1}$, which although consisting of four variables, actually depends on only two variables $A_{1N}$ and $B_{1N}$ (see Equation 7.12) of the outermost layer. The fourth subequation of Equation A7.2.2 gives the relationship between $A_{1N}$ and $B_{1N}$.

## Solid-Fluid Boundary

The case of the solid-fluid boundary is essentially the mirror image of the previous case. Keep with the above notations, the solid layer is now the $n$th layer, and the fluid layer the $n+1$th. The boundary remains at $r = r_n$. The boundary conditions are the same as those given in Equation A7.2.1. At the boundary, we have, in the fluid,

$$\begin{pmatrix} u_{n+1} \\ \sigma_{n+1} \end{pmatrix} = \begin{pmatrix} D_{n+1_{11}} & D_{n+1_{12}} \\ D_{n+1_{31}} & D_{n+1_{32}} \end{pmatrix} \begin{pmatrix} A_{1n+1} \\ A_{1n+1} \end{pmatrix} \qquad A7.2.5$$

and in the solid,

$$\begin{pmatrix} A_{1n} \\ A_{2n} \\ iB_{1n} \\ iB_{2n} \end{pmatrix} = \begin{pmatrix} D_{n11}^{-1} & D_{n13}^{-1} \\ D_{n23}^{-1} & D_{n23}^{-1} \\ D_{n31}^{-1} & D_{n33}^{-1} \\ D_{n41}^{-1} & D_{n43}^{-1} \end{pmatrix} \begin{pmatrix} u_n \\ \sigma_n \end{pmatrix} \qquad A7.2.6$$

From the previous section, we can see how, in general, $A_{1n+1}$ and $A_{2n+1}$ can be expressed in terms of $A_{1N}$ and $B_{1N}$ of the outermost layer. Making use of the boundary condition Equation A7.2.1, $a_n$ can then also be expressed in terms of $A_{1N}$ and $B_{1N}$. The vanishing of the shear stress $\tau_n$ at the boundary providing an additional relationship between the components of $a_n$, reducing the number of unknowns to three, two of which are $A_{1N}$ and $B_{1N}$, with the third to be expressed in terms of $A_{21}$, the coefficient at the center of the borehole.

# 8

# Acoustic Waveforms in Porous and Permeable Formations

According to the continuum theory of elastic solids, the propagation of seismic waves is determined by seismic velocities, densities, and boundary conditions in conjunction with the governing wave equations. However, real rocks are a mixture of minerals and naturally occurring fluids in openings between mineral grains that only approximate a continuous media on a large-scale average. One approach in relating the fine-scale structure of geological formations to acoustic waveforms is correlation between bulk seismic velocities and volumetric composition of rocks. The Wyllie transit time equation could be expanded into the more general formula:

$$\frac{1}{V} = \sum_{i=1}^{N} \frac{C_i}{V_i} \qquad \sum_{i=1}^{N} C_i = 1 \qquad 8.1$$

where $C_i$ is the volumetric fraction of the mineral component with the velocity $V_i$. Equation 2.1 for a water saturated, porous sandstone might be regarded as a "two-mineral" formation composed of quartz and water. This approach to the interpretation of the acoustic properties of rocks does not provide a very sensitive or unique method for characterizing the fluid producing properties of sediments. In some sedimentary reservoirs where lithologies are well known or a great deal of additional information is available, acoustic transit time logs can be related to bulk porosity. In more general cases, the increases in measured transit time produced by fluid-filled porosity cannot be distinguished from transit time increases produced by an increasing fraction of minerals with slower velocities. The most familiar example is the occurrence of shale in sandstone reservoirs, where natural gamma or spontaneous potential logs are used to separate transit time increases caused by shale from those related to porosity. In more general situations where more than one clay mineral may be present, or where there are not enough independent geophysical logs to distinguish all lithologies, measured transit times cannot be used to infer porosity. Moreover, in those situations where porosity can be measured *in situ,* the possibility of inferring permeability from borehole waveform logs is of great interest to the reservoir engineer.

## 8.1    PROPERTIES OF POROUS AND PERMEABLE ROCKS

Many researchers have recognized that acoustic propagation in porous rocks is sensitive to the permeability of the rock matrix. Characterization of this dependence requires a more complete and detailed formulation of the elastic properties of rocks than that given in Chapters 2, 3, and 4. The most important consideration in characterizing wave propagation through permeable formations is that at least some of the saturating fluid is free to move within the porous matrix. Viscous dissipation and inertial coupling in the relative motion between pore fluid and mineral framework introduce important dynamic effects into seismic wave propagation. The porosity of the rock matrix assumed saturated with a single fluid is given as the percent volume of fluid-filled space in the rock. This volume may be entirely unconnected, in which case there can be no continuous flow, and the rock is considered impermeable. Such isolated pores or fluid inclusions are denoted as ineffective porosity because of their inability to transmit flow. In most rocks at least some of these openings are interconnected, and fluid can flow between pores. The volume fraction of pore spaces conducting this flow is known as the effective porosity. The ability of the pore network to transmit fluid is expressed as the permeability of the formation. Permeability or hydraulic conductivity is defined as the amount of fluid transmitted under a unit pressure gradient. This definition is known as Darcy's Law in the hydraulic literature (Bear, 1972):

$$q = \frac{K}{\rho g} P' \qquad\qquad 8.2A$$

where $q$ is the average flow per unit area of formation cross-section, $P'$ is the gradient imposed in the flow direction, $\rho$ is the fluid density, $g$ the acceleration of gravity, and K is the permeability. According to Equation 8.2A, permeability is given in units of velocity. The ability of a given rock sample to transmit fluid depends on the properties of the rock matrix and on the viscosity and density of the saturating fluid. The dependence on fluid properties can be removed by defining an intrinsic permeability, k, which is effectively normalized with respect to the fluid properties (DeWeist, 1969):

$$k = \frac{K \eta}{\rho g} \qquad\qquad 8.2B$$

where $\eta$ is the viscosity of the saturating fluid. The intrinsic permeability has units of square centimeters or darcies, where $1\,D = 0.987 \times 10^{-8}\ cm^2$. In the most general case, permeability depends upon both the direction of flow and the direction of the pressure gradient, and is expressed in tensor form. For almost all borehole applications, both pressure gradient and flow are assumed to be directed along bedding planes in transversely isotropic media. For this reason we will consider the permeability, k, to be a scalar property of the rock similar to porosity and density. Throughout much of the geophysical and hydraulic literature, the pressure gradient is expressed as the gradient in piezometric head (Bear, 1972)

$$h' = \frac{P'}{\rho g} \qquad\qquad 8.3$$

so that

$$q = Kh'$$

The velocity, q, given in Equation 8.2 is the volume averaged discharge per unit area through the rock, known in the hydraulic literature as the darcy velocity. The velocity of flow within the pore spaces is considerably greater to yield this volume-averaged flow. The average fluid velocity within the pores, $q_p$, is given by

$$q_p = \frac{q}{\phi} \qquad\qquad 8.4$$

where $\phi$ is the effective porosity.

## 8.2 WAVE PROPAGATION IN PERMEABLE SOLIDS — THE BIOT MODEL

The extension of synthetic borehole microseismogram theory to porous and permeable formations requires development of a mathematical description of the coupling between seismic waves propagating in the elastic framework and acoustic waves propagating in the pore spaces. The model used in almost every case was developed by Biot (1955, 1956a,b,c), with later refinements by Biot (1962a,b, 1973), Biot and Willis (1957), Stoll (1974), Stoll and Bryan (1970), Berryman (1980a,b,c), Auriault (1980), and Ogushwitz (1985). The original analysis by Biot (1956a) gives a rigorous derivation of the stress-strain relations for a porous and permeable mineral framework saturated with a viscous fluid. Using definitions for the displacements and stresses in the framework and fluid:

$$x_i \;\; = \;\; (r, \theta, z) \qquad\qquad = \text{CYLINDRICAL COORDINATES}$$

$$u_i \;\; = \;\; \left(u, u_\theta, u_z\right) \qquad\qquad = \text{DISPLACEMENT FIELD IN SOLID FRAMEWORK}$$

$$\sigma_{ij} \;\; = \qquad\qquad\qquad\qquad \text{STRESS FIELD IN SOLID FRAMEWORK}$$

$$2e_{ij} \;\; = \;\; \frac{\partial u_i}{\partial x_j} + \frac{\partial u_j}{\partial x_i} \qquad = \text{STRAIN FIELD IN SOLID FRAMEWORK}$$

$$V_i = (V_r, V_\theta, V_z) \qquad \text{DISPLACEMENT FIELD IN FLUID IN PORES}$$

$$s\delta_{ij} = \qquad \qquad \text{STRESS (PRESSURE) FIELD IN THE PORE FLUID}$$

$$2\varepsilon_{ij} = \frac{\partial v_i}{\partial x_j} + \frac{\partial v_j}{\partial x_i} \qquad \text{STRAIN FIELD IN THE PORE FLUID}$$

$$e = e_{ii} \qquad \qquad \varepsilon = \varepsilon_{ii}$$

The stress-strain relations take the form:

$$\sigma_{ij} = 2Ne_{ij} + (Fe + G\varepsilon)\delta_{ij}$$

$$s\delta_{ij} = (Ge + T\varepsilon)\delta_{ij} = -\phi P_f \qquad \qquad 8.6$$

where the various constants $F$, $G$, $T$, and $N$ are defined in terms of the properties of the saturating fluid, the mineral framework, and the bulk mineral from which the framework is constructed:

$$F = \frac{(1-\phi)\left(1 - \phi - \dfrac{\beta_b}{\beta_s}\right)\beta_s + \phi\dfrac{\beta_s}{\beta_f}\beta_b}{1 - \phi - \dfrac{\beta_b}{\beta_s} + \phi\dfrac{\beta_s}{\beta_f}}$$

$$G = \frac{\left(1 - \phi - \dfrac{\beta_b}{\beta_s}\right)\phi\beta_s}{1 - \phi - \dfrac{\beta_b}{\beta_s} + \phi\dfrac{\beta_s}{\beta_f}}$$

$$T = \frac{\phi^2 \beta_s}{1 - \phi - \dfrac{\beta_b}{\beta_s} + \phi\dfrac{\beta_s}{\beta_f}}$$

$$N = \mu_b$$

$$8.6A$$

where $\phi$ is the porosity, $\beta$ is the bulk modulus, and $\mu$ is the shear modulus; the subscripts $s$, $f$, and $b$ denote properties of the mineral, fluid, and porous framework. The bulk modulus of the mineral and fluid, and the shear modulus of the mineral can be expressed in terms of densities and seismic velocities in the form (White, 1983)

$$\beta_s = \rho_s\left(V_p^2 - \frac{4}{3}V_s^2\right), \qquad \mu_s = \rho_s V_s^2 \qquad \beta_f = \rho_f V_f^2 \qquad 8.7A$$

Framework bulk modulus is much more difficult to estimate and would generally require a detailed analysis of the structure of the individual pores in order to predict how they would deform under a given stress field. However, it is assumed that framework bulk modulus decreases with porosity. Most authors (Geertsma, 1957; Seeburger and Nur, 1984) assume a linear relationship between framework bulk modulus and shear strength and porosity. For the purposes of computing microseismograms, we assume a relationship of the form (Schmitt et al., 1988a)

$$\beta_b = (1-\phi)\beta_s$$
$$N = \mu_b = (1-\phi)\mu_s$$

8.7B

Using this relation between stress and strain in the porous solid, the equilibrium condition in the interior of the solid becomes

$$\frac{\partial \sigma_{ij}}{\partial x_j} = \rho_{11}\frac{\partial^2 u_i}{\partial t^2} + \rho_{12}\frac{\partial^2 v_i}{\partial t^2} + b\left(\frac{\partial u_i}{\partial t} - \frac{\partial v_i}{\partial t}\right)$$

$$\frac{\partial s}{\partial x_i} = \rho_{12}\frac{\partial^2 v_i}{\partial t^2} + \rho_{22}\frac{\partial^2 u_i}{\partial t^2} - b\left(\frac{\partial u_i}{\partial t} - \frac{\partial v_i}{\partial t}\right)$$

8.8

where $\rho_{11}$, $\rho_{12}$, and $\rho_{22}$ are mass balance coefficients giving the inertial coupling between fluid and framework motions. $b(\omega)$ is a function of frequency that couples viscous forces in the fluid to the surrounding framework. This definition indicates that $b(\omega)$ is related to the framework permeability.

The mass balance and viscous coupling coefficients remain undefined in Equation 8.8 and must be determined from specific models of porous media. However, in the case of steady fluid flow through a stationary solid, Equation 8.8 reduces to

$$-\phi\nabla P = b(\omega)\frac{\partial v_i}{\partial t} \quad \text{or} \quad \lim_{\omega \to 0} b(\omega) = \frac{\rho g \phi}{K}$$

8.9

by analogy with Darcy's Law. In general, we relate the constant $b(\omega)$ to a complex permeability function (Biot, 1956a; Auriault et al., 1985; Charlaix et al., 1988):

$$b(\omega) = \frac{\rho g \phi}{K(\omega)} = H_R(\omega) + iH_I(\omega)$$

8.10

For the more general case of a uniformly moving solid with no fluid movement, the stress divergence terms in Equation 8.8 vanish and the equations of motion give relations for the mass balance coefficients:

$$\rho_{22} = \frac{\phi^2}{\omega} H_I, \quad \rho_{12} = \phi\rho_s - \rho_{22} \quad \rho_{11} = \phi\rho_f - \rho_{12} \quad b(\omega) = \frac{\phi^2}{\omega} H_R \quad 8.11$$

Following Biot (1956a, b), we use a porous media model composed of a large number of parallel cylindrical tubes of radius $a$. This model gives a relation for the complex permeability function of the form (Schmitt et al., 1988a)

$$K(\omega) = -\frac{\phi}{i\omega\rho_f} \frac{J_2\left(ia\sqrt{\dfrac{i\rho_f\omega}{\eta}}\right)}{J_0\left(ia\sqrt{\dfrac{i\rho_f\omega}{\eta}}\right)} \qquad 8.12$$

The pore radius $a$ in the model is given by the constraint that the low frequency limit of the model match Darcy's Law (Auriault et al., 1985)

$$a^2 = 8\frac{K}{\phi} \qquad 8.13$$

Note that this result is equivalent to the expression given by Biot (1956a) with the structural factor of 8 used by subsequent authors (Rosenbaum, 1974; Bedford et al., 1984). This relation also gives low and high frequency limits for the inertial coupling coefficients

$$\lim_{\omega\to 0} \rho_{22}(\omega) = \frac{4}{3}\rho_f\phi$$
$$\lim_{\omega\to\infty} \rho_{22}(\omega) = \rho_f\phi \qquad 8.14$$

All of the expressions given in Equations 8.8 through 8.14 define the coupled equations for displacements in the framework and fluid equivalent to the single equilibrium equation for an isotropic elastic solid. Biot (1956a,b) and subsequent authors assumed wave solutions to these Equations of the form

$$\phi^b = \left\{A_1{}^b K_0\left(m_{pb}r\right) + A_2{}^b I_0\left(m_{pb}r\right)\right\}e^{i(\kappa z - \omega t)}$$
$$\phi^f = \left\{A_1{}^f K_0\left(m_{pf}r\right) + A_2{}^f I_0\left(m_{pf}r\right)\right\}e^{i(\kappa z - \omega t)}$$
$$\psi = \left\{B_1{}^s K_1\left(m_{sb}r\right) + B_2{}^s I_1\left(m_{sb}r\right)\right\}e^{i(\kappa z - \omega t)} \qquad 8.15$$

$$\zeta = (0, \psi, 0), \quad m_{pb}^{\ 2} = \kappa^2 - \frac{\omega^2}{V_p^{\ 2}} \quad m_{sb}^{\ 2} = \kappa^2 - \frac{\omega^2}{V_s^{\ 2}} \quad m_{pf}^{\ 2} = \kappa^2 - \frac{\omega^2}{V_f^{\ 2}}$$

where $\phi$ and $\zeta$ are scalar and vector potentials for displacements in the composite material composed of solid framework and pore fluid. After some manipulation it is shown that Equation 8.8 yields solutions which satisfy the Helmholtz equation of the form

$$\phi^T = \phi^f + \phi^b$$

$$\nabla^2 \phi^T - \frac{\omega^2}{V_p^{\ 2}} \phi^T = 0 \qquad \nabla^2 \psi^T - \frac{\omega^2}{V_p^{\ 2}} \psi^T = 0$$

and

$$\phi^f = C_1 \phi_1^{\ T} + C_2 \phi_2^{\ T}$$

$$C_i = \frac{1}{V_{pi}^{\ 2}} \frac{(F+2N)T - G^2}{G_{\gamma 22} - T_{\gamma 12}} + \frac{T_{\gamma 11} - G_{\gamma 12}}{G_{\gamma 22} - T_{\gamma 12}}$$

$$\gamma_{12} = \rho_{12} - \frac{ib}{\omega} \qquad \gamma_{11} = \rho_{11} - \frac{ib}{\omega} \qquad \gamma_{22} = \rho_{22} - \frac{ib}{\omega} \qquad 8.16$$

There are two dilatational wave velocities given by the two roots of the equation:

$$V_p^{\ 4}\left(\gamma_{22}\gamma_{11} - \gamma_{12}^{\ 2}\right) - V_p^{\ 2}\left([F+2N]\gamma_{22} + T_{\gamma 11} - 2G_{\gamma 12}\right) + (F+2N)T - G^2 = 0$$

$$8.17$$

where $F$, $N$, $G$, and $T$ are given by Equation 8.7, and the rotational (shear) wave velocity is given by

$$V_s^{\ 2} = \frac{N}{\left(\gamma_{11} - \frac{\gamma_{12}^{\ 2}}{\gamma_{22}}\right)} \qquad 8.18$$

Comparison of the wave propagation in the Biot solution given by Equations 8.17 and 8.18 with the solution for an isotropic elastic solid indicates that the compressional and shear body waves for the latter have become three body waves, two of the dilatational type, and a third closely analogous with the simple shear body wave. However, the two dilatational waves are not simply uncoupled waves propagating in the fluid and solid respectively. Instead, one mode represents the in-phase vibration of the composite fluid/solid system, while the second represents motion where the framework and fluid vibrate out of phase (Biot, 1956a; Johnston et al., 1982). In that sense, the first mode represents wave propagation similar to the conventional compressional wave in a simple elastic solid, whereas the second mode is an entirely new form of wave propagation not present in the elastic case, denoted as a "type II" dilatational wave by Biot (1956a). The third, rotational mode is clearly analogous to the shear body wave in the elastic solid modified by the inertial coupling terms in Equation 8.8.

Equations 8.17 and 8.18 indicate that the effects of formation porosity and permeability on seismic wave propagation are a function of frequency. Inertial coupling and viscous effects increase linearly with frequency at low frequencies. However, at greater frequencies, viscous effects become confined to a narrow boundary layer, so that these effects eventually decrease with increasing frequency. The maxima in viscous and inertial coupling effects are associated with a specific range in the ratio of inertial to viscous forces in the pore spaces. When this ratio approaches unity, viscous effects are maximized and wave attenuation is greatest. Biot (1956b) defined a critical frequency:

$$f_c = \frac{\omega_c}{2\pi} = \frac{\phi\eta}{2\pi K \rho_f} \qquad\qquad 8.19$$

at which attenuation is maximized in a permeable solid. White (1983) used the various fluid and framework properties for porous solids to compute values of $f_c$ ranging from less than 1 kHz (permeable sand and gravel with $\kappa = 300$ D) to more than 10 kHz (consolidated sandstones with $\kappa = 10$ mD).

Variations of the fluid inertial ($\rho_{22}$) and viscous ($b$) coupling terms over the frequency range including typical acoustic logging frequencies are shown in Figure 8.1. The effects of viscous coupling between fluid and framework become important at logging frequencies (10 to 50 kHz) when formation permeability becomes greater than 1.0 D in a water-saturated formation. Group and phase velocities for the two dilatational and shear waves are given in Figure 8.2. The effects of the properties of the saturating fluid on frequency dependence are illustrated in Figure 8.3 for the case where formation permeability is moderate (0.2 D). The lower densities of natural gas and oil shift the effects of formation permeability toward slightly lower frequencies in accordance with Equation 8.19, with the shift in approximate proportion to the density differences between oil, gas, and water.

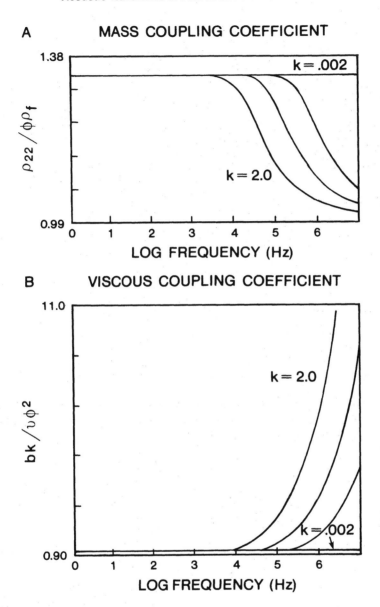

FIGURE 8.1.    Values of (A) the mass coupling coefficient ($p_{22}$) and (B) viscous coupling coefficient (b) in the Biot model computed for permeabilities ranging from 0.002 to 2.0 D; sandstone lithology with $\phi = 19\%$.

## 8.3    MICROSEISMOGRAMS IN PERMEABLE FORMATIONS

Although the derivation of the Biot formulation for seismic propagation in an elastic solid appears very complicated, the procedures for computation of the

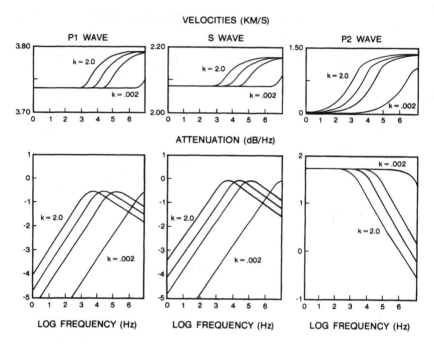

FIGURE 8.2.    Values for the phase and group velocities and attenuation of the two dilatational and shear waves in a sandstone with permeability varying from 0.002 to 2.0 D.

microseismograms are only slightly more complex than in the case of the isotropic elastic medium. In the Biot formulation, we consider a superposition of three wave solutions of the form

$$\phi_i = \left\{ A_1^{(i)} K_0\left(m_{pi}r\right) + A_2^{(i)} I_0\left(m_{pi}r\right) \right\} e^{i(\kappa z - \omega t)}$$

$$\psi = \left\{ B_1 K_0\left(m_s r\right) + B_2 I_0\left(m_s r\right) \right\} e^{i(\kappa z - \omega t)}$$

8.20A

$$m_{pi}^{\ 2} = \kappa^2 - \frac{\omega^2}{V_{pi}^{\ 2}} \qquad m_s^{\ 2} = \kappa^2 - \frac{\omega^2}{V_s^{\ 2}}$$

where the coefficients $A_2^{(i)}$ and $B_2$ are set equal to zero to satisfy radiation boundary conditions, and $A_1^{(i)}$ and $B_1$ are functions of frequency and wavenumber to be determined by the boundary and radiation conditions. The two dilatational potentials, $\phi_i$, represent the two coupled vibrations of framework and pore fluid defined as Equation 8.16. These solutions apply to the permeable formation surrounding the borehole. Within the borehole fluid we have the single dilatational solution of the form

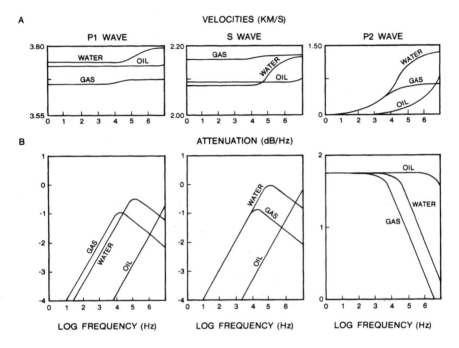

FIGURE 8.3.    Values for the (A) phase and group velocities and (B) attenuation of the two dilatational and shear waves in a sandstone with a permeability of 0.2 D, and saturated with oil, water, or gas.

$$\phi_f = \left\{ D_1 I_0\left(m_f r\right) + D_2 K_0\left(m_f r\right) \right\} e^{i(\kappa z - \omega t)}$$

$$m_f^{\,2} = \kappa^2 - \frac{\omega^2}{V_f^{\,2}}$$

8.20B

For the sake of simplicity we assume that waves are excited in a fluid-filled borehole without logging tool, so the modified Bessel function of the second kind is removed from the solution (i.e., $D_2 = 0$).

Displacements and stresses in the framework and pore fluid are given by

$$u_r = \sum_{i=1}^{2} \frac{\partial \phi_i}{\partial r} - \frac{\partial \psi}{\partial r}$$

$$u_z = \sum_{i=1}^{2} \frac{\partial \phi_i}{\partial z} + \frac{\psi}{r} + \frac{\partial \psi}{\partial r}$$

$$v_r = \sum_{i=1}^{2} A_i \frac{\partial \phi_i}{\partial r} - \frac{\gamma_{12}}{\gamma_{22}} \frac{\partial \psi}{\partial r}$$

$$\sigma_{rr} + s = 2N \left\{ \sum_{i=1}^{2} \frac{\partial^2 \phi_i}{\partial r^2} - \frac{\partial^2 \psi}{\partial r \partial z} \right\} - \sum_{i=1}^{2} \frac{\omega^2}{V_p^2} \left\{ F + G + C_i (G + T) \right\} \phi_i$$

$$\sigma_{rz} = -\frac{\omega^2}{V_\beta^2} N\psi + 2N \left\{ \sum_{i=1}^{2} \frac{\partial^2 \phi_i}{\partial r \partial z} - \frac{\partial^2 \psi}{\partial z^2} \right\} \qquad \qquad 8.21$$

where the parameters $C_i$ are derived from the expression for the scalar fluid potential $\phi_f$ in terms of the solid frame $\phi^T$ potential in Equation 8.16. The solution is completed by invoking the boundary conditions at the borehole wall:

1. Continuity of fluid flow
2. Continuity of normal stress
3. Continuity of axial shear stress
4. Continuity of pressure

Fitting the general solution with these conditions results in a set of four coupled algebraic equations for the evaluation of the four coefficients ($A_1^{(1)}$, $A_1^{(2)}$, $B_1$, and $D_1$) which remain after enforcement of the radiation conditions. The expression for each of these coefficients is given in Appendix 8.1.

In one of the first applications of the Biot theory to acoustic logging, Rosenbaum (1974) modified the analysis to include the possible effects of mudcake on the borehole wall. The mudcake effect was approximated by including an impedance factor representing an arbitrary reduction in the continuity between radial flow in the borehole and flow in the pore spaces. Schmitt et al. (1988a) questioned the physical correctness of such a representation. We consider the extreme limit of the Rosenbaum (1974) formulation by imposing an impermeable barrier at the borehole wall. In that case, the first boundary condition, continuity of radial flow at the wall, becomes

$$u_r(\text{fluid}) = u_r(\text{solid}) \qquad \qquad 8.22$$

The matrix formulation of the equations for the coefficients in Equation 8.20 for the impermeable wall case is given in Appendix 8.1.

Solutions of the coupled linear equations given in Appendix 8.1 for the Fourier coefficients, multiplication by source spectrum, and then inversion in wavenumber and frequency space, produces the general solution to borehole forcing in fluid-saturated, porous media. We illustrate some of the most significant effects of permeability on seismic propagation in porous formations by computing microseismograms for a typical coarse sandstone, the Berea Sandstone model used by Rosenbaum (1974). We also compute the microseismograms

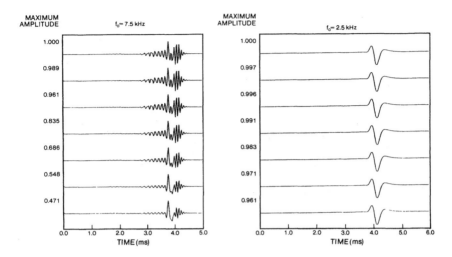

FIGURE 8.4.    Synthetic microseismograms for porous sandstone of variable permeability ranging from impermeable elastic solid to 1.5 D; computed at source frequencies of (A) 2.5 kHz and (B) 7.5 kHz; impermeable mudcake.

for an equivalent elastic formation, where the seismic velocities of the elastic solid are given as those at the low frequency limit for the permeable solid, and the formation density is given by

$$\rho = \phi\rho_f + (1-\phi)\rho_s \qquad\qquad 8.23$$

The effects of increasing formation permeability are illustrated in Figure 8.4 for the impermeable wall solution and in Figure 8.5 for the permeable wall solution. In each case formation permeabilities vary from 0 (equivalent elastic formation) to 1.5 D, with microseismograms given for centerband frequencies of 7.5 and 2.5 kHz. The pair of source frequencies used for the calculations in Figures 8.4 and 8.5 have been selected to correspond to practical logging situations where the source excites the lowest pseudo-Rayleigh mode near low frequency cutoff (7.5 kHz at the modeled radius) or at lower frequencies for good tube wave excitation (2.5 kHz). In all computations, the fluid pressure is evaluated along the borehole axis. Figure 8.4 shows a small increase in attenuation of the microseismogram for increasing permeability. The greatest attenuation is associated with the head waves, which travel within the permeable formation. In Figure 8.5, the higher frequencies are associated with much more attenuation, and the head waves appear enhanced because of the normalized representation of the computed microseismograms. That is, the entire waveform is attenuated for increasing permeability, but the tube wave and pseudo-Rayleigh modes are much more attenuated than the head waves. These results serve to illustrate the great sensitivity of the trapped modes (tube waves and pseudo-Rayleigh waves) to borehole wall permeability as demonstrated by Rosenbaum

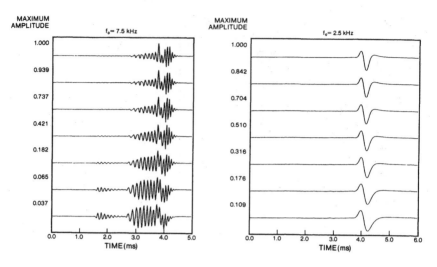

FIGURE 8.5.    Synthetic microseismograms for porous sandstone of variable permeability ranging from impermeable elastic solid to 1.5 D; computed at source frequencies of (A) 2.5 kHz and (B) 7.5 kHz; no mudcake.

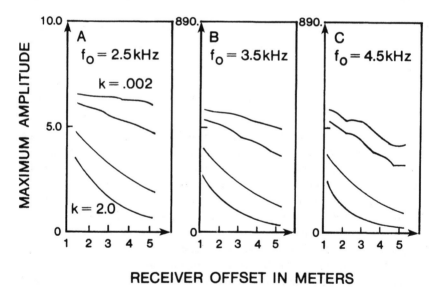

RECEIVER OFFSET IN METERS

FIGURE 8.6.    Variation of maximum amplitude in synthetic microseismograms with increasing receiver offset for a porous sandstone; permeability varies from 0.002 to 2.0 D for source frequencies of (A) 2.5 kHz, (B) 3.5 kHz, and (C) 4.5 kHz.

(1974), Hsui et al. (1985), Cheng et al. (1987), Burns (1989), and Winkler et al. (1989). The increasing attenuation of the microseismograms with increasing permeability is summarized in Figure 8.6, where maximum synthetic microseismogram amplitude is given for increasing source to receiver offset for the same series of permeabilities given in Figures 8.4 and 8.5. These results duplicate those for the fully permeable borehole wall case given by Rosenbaum (1974).

The increased attenuation of the trapped modes indicates that there will be a significant effect on the appearance of waveforms caused by the preferential attenuation of guided modes when mudcake does not seal the borehole wall. Furthermore, the increased effect of attenuation associated with permeability with increasing frequency indicates that this effect will be nonuniform across the range of frequencies associated with acoustic logging. This effect is illustrated in Figure 8.7 where microseismograms for a permeable sandstone (Berea model with $k = 2$ mD) are plotted for frequencies ranging from 1.5 to 7.5 kHz. The low-frequency waveforms are dominated by the tube wave, but the tube wave is no longer apparent at frequencies greater than 5 kHz and the pseudo-Rayleigh mode is never apparent at all. All of these results and conclusions parallel the results given by Rosenbaum (1974). At the time when the Rosenbaum paper appeared, these results were taken as an indication that the presence of mudcake in boreholes penetrating permeable formations would make tube wave attenuation associated with formation permeability difficult to detect. However, even in the limit of a completely impermeable barrier at the borehole wall, the permeability within the formation still allows for slightly increased attenuation of waves measured in the borehole fluid. It is also unlikely that mudcake forms a perfect barrier to the borehole fluid. Nevertheless, mud properties are controlled during drilling for a variety of purposes. Often this manipulation includes fluid loss control, so that an effective barrier is being set up and the effect of mudcake on acoustic waveforms recorded in the borehole needs to be considered.

When wells have been developed in order to remove mudcake, or when conditions permit drilling without mud additives to impede fluid loss, forward modeling of waveforms using the Biot theory can be a useful method for estimating formation permeability by means of tube wave and pseudo-Rayleigh mode attenuation. In practice this could vary from qualitative identification of the depths where attenuation attributed to formation permeability is likely to be the greatest, to precise matching of microseismograms with borehole wave-forms. One especially useful approach might include the ratio of maximum amplitudes in the compressional arrivals to the tube wave or pseudo-Rayleigh arrivals. The slight dispersion of the tube wave mode and the ability to filter tube waves out of the waveforms on the basis of their frequency content indicate that tube wave amplitude is superior to other amplitude data in estimating formation permeability.

## 8.4  EFFECTS OF PORE PROPERTIES ON MICROSEISMOGRAMS IN POROUS FORMATIONS

The microseismograms illustrated in Figures 8.4, 8.5, and 8.7 were calculated using a model for pore structure based on the array of cylindrical tubes of radius

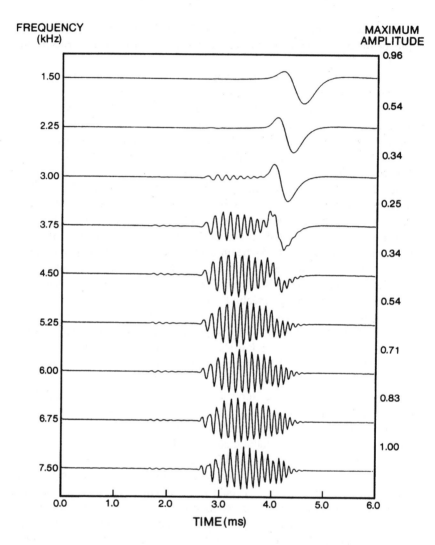

FIGURE 8.7.    Synthetic microseismograms for permeable sandstone ($k = 0.002$ D) with no mudcake illustrating microseismogram variation with frequency from 1.5 to 7.5 kHz.

a introduced by Biot ( 1956a,b). Other models for pore structure could be used (Brace, 1977; Bedford et al., 1984). How much of a difference does the exact pore model structure make in the interpretation of waveforms in porous media? The pore model affects the calculations through the inertial coupling terms $\rho_{11}$, $\rho_{12}$, and $\rho_{22}$, and through the viscous coupling term $b(\omega)$.

One example of the effects of different pore structure models is illustrated in Figures 8.8 and 8.9, where microseismograms are shown for two different pore structures at various frequencies from 1.5 to 7.5 kHz. The two sets of

FREQUENCY
(kHz)

MAXIMUM
AMPLITUDE

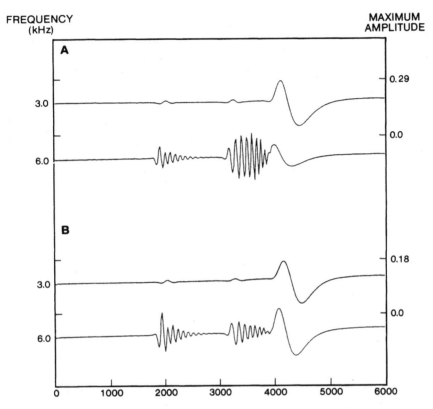

FIGURE 8.8.    Synthetic microseismograms computed for source frequencies of 3.0 and 6.0 kHz for permeable sandstone using (A) parallel tube pore model and (B) tortuous tube pore model.

microseismograms have been calculated for exactly the same sandstone formation, except that the second set has been modeled with a greater pore tortuosity. The increased tortuosity of the pores is expressed as a larger inertial coupling between the fluid and the mineral framework (Schmitt et al., 1988a). The increased inertial coupling makes the second model more effective in extracting energy from the framework, an effect that increases with increasing frequency for $\omega < 2\eta f_c$. This increased attenuation is indicated by the greater attenuation for the second set of microseismograms, corresponding to the larger tortuosity. However, the increase in attenuation is confined almost completely to the pseudo-Rayleigh mode. Comparison of the microseismograms indicates that the tube wave is attenuated by approximately the same amount in both cases. This appears to result from the lower frequency content and reduced sensitivity to formation properties associated with the tube wave. This is another argument for recommending the use of tube wave rather than pseudo-Rayleigh amplitudes in estimating formation permeability.

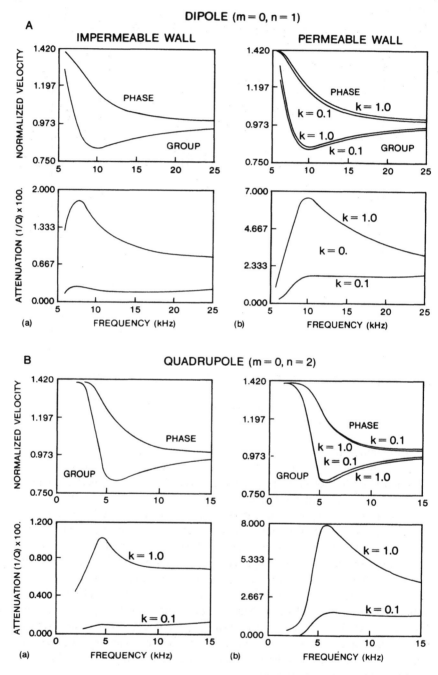

FIGURE 8.9.    Dispersion and attenuation of (A) dipole and (B) quadrupole modes for a porous sandstone computed using three different values of permeability.

## 8.5   MICROSEISMOGRAMS FOR NONSYMMETRIC MODES IN POROUS FORMATIONS

In the previous sections of this chapter, solutions to Equation 8.8 were assumed to be symmetric about the borehole axis. The addition of azimuthal dependence allows the polarized shear waves to be expressed by two potentials, requiring two independent shear solutions:

$$\zeta_1 = (0,0,\psi), \quad \zeta_2 = (0,0,\Gamma)$$

$$v = \sum_{i=1}^{2} \rho_i \nabla\phi_i + \nabla\times\zeta_1 + \nabla\times\nabla\times\zeta_2$$

$$u = \sum_{i=1}^{2} \nabla\phi_i + \nabla\times\zeta_1 + \nabla\times\nabla\times\zeta_2$$

$$\phi^f = \sum_{n=1}^{\infty} D_n I_n\left(m_f r\right)\cos(n\theta)e^{i(\kappa z-\omega t)}$$

$$\phi_i^b = \sum_{i=1}^{2}\sum_{n=1}^{\infty} A_{in} K_n\left(m_{pi} r\right)\cos(n\theta)\, e^{i(\kappa z-\omega t)}$$

$$\psi = \sum_{n=1}^{\infty} B_n K_n\left(m_s r\right)\sin(n\theta)\, e^{i\left(\kappa^2-\omega t\right)} \qquad\qquad 8.24$$

$$\Gamma = \sum_{n=1}^{\infty} E_n K_n\left(m_s r\right)\cos(n\theta)\, e^{i(\kappa z-\omega t)}$$

$$m_f^{\,2} = \kappa^2 - \frac{\omega^2}{V_f^{\,2}} \qquad m_{pi}^{\,2} = \kappa^2 - \frac{\omega^2}{\alpha_i^{\,2}} \qquad m_s^{\,2} = \kappa^2 - \frac{\omega^2}{V_s^{\,2}}$$

where $\alpha_i$ are the two roots to the Equation 8.17 for the two dilatational waves in the porous solid, and $V_s$ is the rotational or shear wave velocity (Equation 8.18) modified by the inertial coupling between pore fluid and mineral framework. These potentials satisfy the governing wave equation. The conditions for the permeable and impermeable borehole walls, along with Equation 8.22 are used for the symmetric case, with the addition of one more condition:

> 5. Continuity of azimuthal shear stress: $\sigma_{z\theta} = 0$      at $r = R$      8.25

For practical purposes, we insert this condition in place of condition 3 making condition 3 condition 4, and condition 4 condition 5. The elements in the resulting set of coupled algebraic equations for the determination of the Fourier coefficients in Equation 8.24 for both permeable and impermable wall cases are given in Appendix 8.2. After the Fourier coefficients have been solved, the pressure response in the borehole can be synthesized by inversion of the integral

4.35 using the displacement potential in the borehole fluid, which has exactly the same mathematical form as in solutions for the elastic solid case of Chapter 4. In the case of permeable formations, the effect of permeability is enforced through matching with the solutions for the permeable solid at the borehole wall.

Porous and permeable formations allow for attenuation of trapped nonsymmetric modes in a manner similar to that for the symmetric modes. In describing these calculations we shall keep to the familiar convention for describing modes: $n$ denotes the azimuthal periodicity, while $m$ denotes the index of the individual modes corresponding to the infinite series for each value of $n$. In particular, the $m = 0$ mode corresponds to the fundamental mode, which becomes a Stoneley interface mode in the high frequency limit.

Velocity dispersion and attenuation of the $n = 0$, $n = 1$, and $n = 2$; $m = 0$ modes are compared for increasing values of permeability in Figure 8.9. In contrast to the result in with symmetrical deformation around the circumference of the borehole, there is only a slight increase in attenuation for the case of permeable borehole wall over the impermeable wall in the figure. Microseismograms for the flexural or dipole ($n = 1$, $m = 0$) mode for 1 and 3 kHz source are illustrated in Figure 8.10. Increasing permeability increases attenuation by nearly the same amount for both the impermeable and the permeable wall cases. Results for the quadrupole mode are similar. The source frequency is much lower than the frequencies at which mode velocity becomes dispersive, so that calculated mode arrivals are very close to that determined from ray tracing using formation shear velocity. This figure was chosen for illustration in Figure 8.10 as typical of those relatively low frequencies that would be useful in shear logging with nonsymmetric sources.

In general, the results obtained using the nonsymmetric solutions Equation 8.24 confirm the results obtained in the symmetric case, with one important exception. This is the almost complete lack of an effect of borehole wall sealing (as with mudcake). The difference results from the ability of a single sealed but deformable layer to inhibit symmetrical fluid flow away from the borehole axis. Borehole wall sealing has little effect on a nonsymmetric motion because such motion does not require a net transfer of fluid across the borehole wall. The permeable formation calculations using the Biot (1956a,b) formulations give results in which the coupling between formation and pore fluid produce small changes in mode velocities and greatly increased attenuation of guided modes. The attenuation effects are best measured by using the amplitude of the fundamental mode (tube wave for $n = 0$; $m = 0$ modes for $n > 0$) where there is low velocity dispersion and frequency filtering can be applied.

## 8.6    INTERPRETATION OF *IN SITU* PERMEABILITY IN POROUS FORMATIONS

The synthetic microseismograms computed for porous and permeable formations indicate a significant dependence of both waveform amplitude and mode velocities on formation porosities. These theoretical predictions can be used to

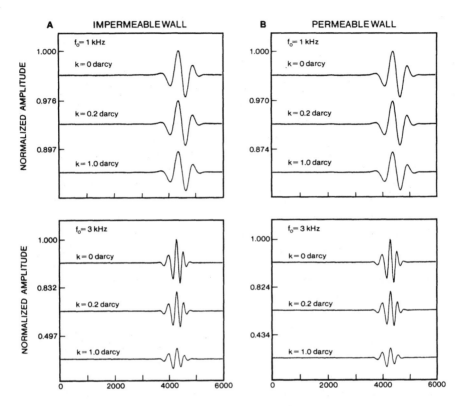

FIGURE 8.10.   Synthetic microseismograms computed using source frequencies of 1 and 3 kHz for a dipole source ($m = 0$, $n = 1$) assuming (A) impermeable borehole wall and (B) permeable borehole wall; arrow denotes direct shear travel time.

evaluate the permeability of rocks adjacent to the borehole when acoustic waveform logs are available and when enough additional information is on hand to separate the effects of lithology and borehole conditions from the effects of permeability. As of the publication of this monograph (1990), relatively few studies have compared predictions based on the Biot (1956a,b) theory with experimental waveforms, and even fewer have attempted to compare synthetic microseismograms with waveforms obtained in permeable formations. Early efforts to compare waveforms obtained in boreholes with the Biot theory were reviewed by White (1983). Although the theory shows that all wave modes are affected by the formation permeability, the microseismograms indicate that the Stoneley or tube wave mode is the most effective indicator of permeability. Recent examples of permeability interpretation using tube wave data for porous formations were given by Cheng et al. (1987) and Burns et al. (1988). Winkler et al. (1989) used a laboratory scale model of a borehole penetrating a permeable formation to investigate the relationship between tube wave amplitude and permeablility. Various aspects of the theoretical interpretation of permeability effects on tube wave propagation were described by Pascal (1986) and Norris (1989).

The most direct approach to the interpretation of permeability using tube wave data is the correlation of tube wave velocity or amplitude with formation permeability. This method circumvents the need for a detailed description of formation properties required to match synthetic microseismograms with waveforms, but requires a reliable independent measure of permeability with which to calibrate the waveform data. The most common form of permeability data are the results of core permeability tests. These data cannot be directly applied to the calibration of waveform data because of the different sample volumes associated with tube waves (about 1 m) and core permeability measurements (1 to 2 cm). One also needs to assume that there will be a difference between the depth scale given on the log and the depths assigned to core samples. These problems can be resolved by smoothing the vertical distribution given by a large number of core permeabilities, and then correlating tube wave slowness (inverse velocity) or tube wave amplitude with the permeability data. In general, a small depth offset will be needed to maximize the correlation. Once the correlation has been completed and the results are shown to yield a statistically significant correlation, the data can be used to calibrate waveform amplitude of tube wave slowness in terms of formation permeability.

Burns (1989) and Burns et al. (1986, 1988) demonstrated that both tube wave slowness and amplitude can be correlated with permeability using data given by Williams et al. (1984). Comparison of tube wave attenuation (ratio of amplitudes at two receivers) and slowness (interval transit time between the same two receivers) with measured core permeabilities for sandstone and limestone using the data given by Williams et al. (1984) are illustrated in Figure 8.11. The data show the expected correlation between amplitude and slowness and measured permeability values. However, a major complication in estimating formation permeability using tube wave slowness is the dependence of tube wave velocity on lithologic factors such as shear strength that are effectively independent of permeability. Burns et al. (1988) resolved this problem by correcting for the tube wave slowness variations predicted from the independently measured shear velocity and then associating the residual slowness variations with formation permeability. The results for the shear velocity corrected permeability interpretations are plotted for both the sandstone and limestone data in Figure 8.12. The correlation between permeability and tube wave slowness appears to define a single line. Analogous results are given by the laboratory model study of the correlation between tube wave amplitude and permeability (Winkler et al., 1989). Their results indicate that increases in tube wave slowness of as much as 10% are associated with relatively large formation permeabilities (greater than 1 D), and that the lower limit of permeability resolution is about 10 mD.

Burns et al. (1988) also reported attempts to directly model the tube wave attenuation in the Williams et al. (1984) data using synthetic microseismogram calculations based on the Biot (1956a,b) formulation. The results are relatively successful for the limestone data using a zero value for borehole wall impedance, as given by Rosenbaum (1974), and analogous to the permeable borehole wall condition in this chapter. However, the microseismogram calculations appear to

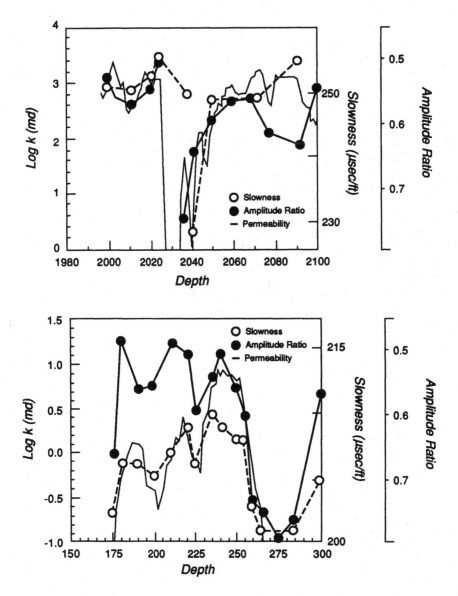

FIGURE 8.11    Correlation of tube wave amplitude ratio and tube wave slowness with measured permeability form core samples for (top) sandstone and (bottom) limestone from Burns et al. (1988) using data given by Williams et al. (1984).

overpredict the tube wave attenuation for the sandstone data. Burns (1989) reported an improved fit of the microseismogram predictions using a value for borehole wall permeability intermediate between the permeable and impermeable limits used earlier in this chapter. This result could represent the effect of filtercake build-up on the borehole wall, but may also result from uncertainty

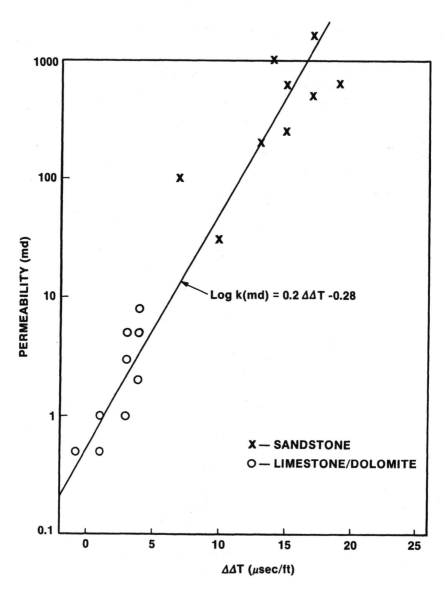

FIGURE 8.12 Correlation of tube wave slowness corrected for lithology with measured core permeabilities for the data given in Figure 8.11.

encountered in specifying the mineral skeleton or framework properties for the model calculations. We hope that in the future waveform logs obtained with nonsymmetric transducers where waveform amplitudes are not so strongly affected by borehole wall permeability will clarify whether the results reported by Burns et al. (1988) and Burns (1989) result from mudcake effects or the inability to specify mineral framework properties for the microseismogram model.

# Appendix 8.1

## Boundary Condition Matrix for Axisymmetric Modes in a Fluid-Filled Borehole in a Porous Solid

The four conditions at the borehole wall $(r = R)$ for a fluid-filled borehole in a porous solid are based on the continuity of radial displacement, radial stress, and pressure, and the vanishing of axial shear stress. The fourth boundary condition (compare to Equation 4.31) arises because both formation stress and pore fluid pressure must match the pressure in the borehole fluid. This gives a set of four equations of the form

$$
\begin{array}{l}
\text{CONTINUITY OF RADIAL DISPLACEMENT} \\
\text{CONTINUITY OF NORMAL STRESS} \\
\text{VANISHING OF AXIAL SHEAR STRESS} \\
\text{CONTINUITY OF PORE PRESSURE}
\end{array}
\begin{pmatrix}
\theta_{11} & \theta_{12} & \theta_{13} & \theta_{14} \\
\theta_{21} & \theta_{22} & \theta_{23} & \theta_{24} \\
0 & \theta_{32} & \theta_{33} & \theta_{34} \\
\theta_{41} & \theta_{42} & \theta_{43} & 0
\end{pmatrix}
\begin{pmatrix}
D_1 \\
A_{1(2)}^{(1)} \\
A_1 \\
B_1
\end{pmatrix}
=
\begin{pmatrix}
S_1 \\
S_2 \\
0 \\
S_2
\end{pmatrix}
\qquad A8.1
$$

where the column vector $(S_1, S_2, 0, S_2)$ is given by the pressure field

$$
S_1 = -m_f K_1\left(m_f R\right)
$$
$$
S_2 = -\rho_f \omega^2 K_0\left(m_f R\right) \qquad A8.2
$$

produced by a point source at the borehole wall.

The nonzero elements of the $\theta_{ij}$ are

$$
\theta_{11} = -m_f I_1\left(m_f R\right)
$$
$$
\theta_{12} = m_{p1} K_1\left(m_{p1} R\right)\left[1 - \phi\left(1 - C_1\right)\right]
$$
$$
\theta_{13} = m_{p2} K_1\left(m_{p2} R\right)\left[1 - \phi\left(1 - C_2\right)\right]
$$
$$
\theta_{14} = i\kappa K_1\left(m_s R\right)\left[1 - \phi\left(1 - \chi\right)\right]
$$
$$
\theta_{21} = \rho_f \omega^2 I_0\left(m_f R\right)
$$
$$
\theta_{22} = -2Nm_{p1}^{2}\left[K_0\left(m_{p1}R\right) + \frac{1}{m_{p1}R}K_1\left(m_{p1}R\right)\right] + \frac{\omega^2}{V_{p1}^{2}}\left[F + G + (G+T)C_1\right]K_0\left(m_{p1}R\right)
$$
$$
\theta_{23} = -2Nm_{p2}^{2}\left[K_0\left(m_{p2}R\right) + \frac{1}{m_{p2}R}K_1\left(m_{p2}R\right)\right] + \frac{\omega^2}{V_{p2}^{2}}\left[F + G + (G+T)C_2\right]K_0\left(m_{p2}R\right)
$$
$$
\theta_{24} = -2\kappa m_s\left[K_0\left(m_s R\right) + \frac{1}{m_s R}K_1\left(m_s R\right)\right] \qquad A8.3
$$

$$\theta_{32} = -2iN\kappa m_{p1} K_1\left(m_{p1}R\right)$$

$$\theta_{33} = -2iN\kappa m_{p2} K_1\left(m_{p2}R\right)$$

$$\theta_{34} = N\left[2\kappa^2 - \frac{\omega^2}{V_s^2} K_1\left(m_s R\right)\right]$$

$$\theta_{41} = \rho_f \omega^2 I_0\left(m_f R\right)$$

$$\theta_{42} = \frac{\omega^2}{\phi V_{p1}^{\,2}}\left[G+TC_1\right]K_0\left(m_{p1}R\right)$$

$$\theta_{43} = \frac{\omega^2}{\phi V_{p2}^{\,2}}\left[G+TC_2\right]K_0\left(m_{p2}R\right)$$

Where the constants $F, G, T, N$, and $C_i$ are defined in Equations 8.7 and 8.16, and

$$\chi = -\frac{\gamma_{12}}{\gamma_{22}}$$

In situations where the borehole wall is assumed to be impermeable, the boundary conditions assume the form

$$\begin{pmatrix} \theta_{11}\theta_{12}\theta_{13}\theta_{14} \\ \theta_{21}\theta_{22}\theta_{23}\theta_{24} \\ 0\ \ \theta_{32}\theta_{33}\theta_{34} \\ 0\ \ \theta_{42}\theta_{43}\theta_{44} \end{pmatrix} \begin{pmatrix} D_{1(1)} \\ A_{1(2)} \\ A_1 \\ B_1 \end{pmatrix} = \begin{pmatrix} S_1 \\ S_2 \\ 0 \\ 0 \end{pmatrix}$$

All of the coefficients remain the same with the exception of

$$\theta_{13} = m_{p2} K_1\left(m_{p2}R\right)$$

$$\theta_{14} = i\kappa K_1\left(m_s R\right)$$

$$\theta_{42} = m_{p1}\left[C_1 - 1\right]K_1\left(m_{p1}R\right)$$

$$\theta_{43} = m_{p2}\left[C_2 - 1\right]K_1\left(m_{p2}R\right)$$

$$\theta_{44} = i\kappa\left[\chi - 1\right]K_1\left(m_s R\right) \qquad\qquad A8.4$$

# Appendix 8.2

## Boundary Condition Matrix for Nonaxisymmetric Modes in a Fluid-Filled Borehole in a Porous Solid

In generalizing the formulation to include nonaxisymmetric modes, the boundary conditions are expanded to include a fifth condition, the vanishing of azimuthal shear stress (in addition to the axial shear stress) at the borehole wall. In doing so we modify the original conditions by inserting the azimuthal stress condition in the position of condition 3, making the axial stress condition now condition 4, and the continuity of pressure now condition 5:

$$
\begin{matrix}
\text{CONT. OF RADIAL DISP.} \\
\text{CONT. OF NORMAL STRESS} \\
\text{ZERO AZIMUTHAL SHEAR STRESS} \\
\text{ZERO AXIAL SHEAR STRESS} \\
\text{CONT. OF PORE PRESSURE}
\end{matrix}
\quad
\begin{pmatrix}
\theta_{11} & \theta_{12} & \theta_{13} & \theta_{14} & \theta_{15} \\
\theta_{21} & \theta_{22} & \theta_{23} & \theta_{24} & \theta_{25} \\
0 & \theta_{32} & \theta_{33} & \theta_{34} & \theta_{35} \\
0 & \theta_{42} & \theta_{43} & \theta_{44} & \theta_{45} \\
\theta_{51} & \theta_{52} & \theta_{53} & 0 & 0
\end{pmatrix}
\begin{pmatrix}
D_n \\
A_{1n} \\
A_{2n} \\
B_n \\
E_n
\end{pmatrix}
=
\begin{pmatrix}
S_{1n} \\
S_{2n} \\
0 \\
0 \\
S_{2n}
\end{pmatrix}
\qquad A8.5
$$

The source terms are

$$
S_{1n} = \frac{nK_n\left(m_f R\right)}{R} - m_{fK_{n+1}\left(m_f R\right)}
$$

$$
S_{2n} = -\rho_f \omega^2 K_n\left(m_f R\right) \qquad\qquad n \geq 1
$$

$$A4.7$$

The nonzero elements of the $\theta_{ij}$ matrix are

$$
\theta_{11} = \frac{nI\left(m_f R\right)}{R} + m_f I_{n+1}\left(m_f R\right)
$$

$$
\theta_{12} = -\left[\frac{nK_n\left(m_{p1} R\right)}{R} - m_{p1}K_{n+1}\left(m_{p1} R\right)\right]\left[1 - \phi\left(1 - C_1\right)\right]
$$

$$
\theta_{13} = -\left[\frac{nK_n\left(m_{p2} R\right)}{R} - m_{p2}K_{n+1}\left(m_{p2} R\right)\right]\left[1 - \phi\left(1 - C_2\right)\right]
$$

$$
\theta_{14} = -\frac{nK_n\left(m_s R\right)}{R}\left[1 - \phi\left(1 - \chi\right)\right]
$$

$$
\theta_{15} = -\iota\kappa\left[\frac{nK_n\left(m_s R\right)}{R} - m_s K_{n+1}\left(m_s R\right)\right]\left[1 - \phi\left(1 - \chi\right)\right]
$$

$$
\theta_{21} = -\rho_f \omega^2 I_n\left(m_f R\right)
$$

$$\theta_{22} = -2N\left\{ I_n\left(m_{p1}R\right)\left[ m_{p1}^{2} + \frac{n(n-1)}{R^2}\right] - \frac{1}{R}m_{p1}I_{n+1}\left(m_{p1}R\right)\right\} + \frac{\omega^2}{V_{p1}^{2}}\left[F + G + (G+T)C_1\right]I_n\left(m_{p1}R\right)$$

$$\theta_{23} = -2N\left\{ I_n\left(m_{p2}R\right)\left[ m_{p2}^{2} + \frac{n(n-1)}{R^2}\right] - \frac{1}{R}m_{p2}I_{n+1}\left(m_{p2}R\right)\right\} + \frac{\omega^2}{V_{p2}^{2}}\left[F + G + (G+T)C_2\right]I_n\left(m_{p2}R\right)$$

$$\theta_{24} = -2N\left[\frac{n-1}{R^2}\right]K_n\left(m_sR\right) - \frac{nm_s}{R}K_{n+1}\left(m_sR\right)$$

$$\theta_{25} = -2iN\kappa\left\{ K_n\left(m_sR\right)\left[ m_s^{2} + \frac{n(n-1)}{R^2}\right] + \frac{1}{R}m_sK_{n+1}\left(m_sR\right)\right\}$$

$$\theta_{32} = 2N\left[\frac{n(n-1)}{R^2}K_n\left(m_{p1}R\right) - \frac{n}{R}K_{n+1}\left(m_{p1}R\right)\right]$$

$$\theta_{33} = 2N\left[\frac{n(n-1)}{R^2}K_n\left(m_{p2}R\right) - \frac{n}{R}K_{n+1}\left(m_{p2}R\right)\right]$$

$$\theta_{34} = N\left\{ K_n\left(m_sR\right)\left[ m_s^{2} + \frac{2n(n-1)}{R^2}\right] + \frac{2}{R}m_sK_{n+1}\left(m_sR\right)\right\}$$

$$\theta_{35} = 2iN\kappa\left\{ K_n\left(m_sR\right)\frac{n(n-1)}{R^2} + \frac{n}{R}m_sK_{n+1}\left(m_sR\right)\right\}$$

$$\theta_{42} = -2iN\kappa\left\{ K_n\left(m_{p1}R\right)\frac{n}{R} - m_{p1}K_{n+1}\left(m_{p1}R\right)\right\}$$

$$\theta_{43} = -2iN\kappa\left\{ K_n\left(m_{p2}R\right)\frac{n}{R} - m_{p2}K_{n+1}\left(m_{p2}R\right)\right\}$$

$$\theta_{44} = -\frac{iN\kappa}{R}K_n\left(m_sR\right)$$

$$\theta_{45} = N\left[ 2\kappa^2 - \frac{\omega^2}{V_s^{2}}\right]\left\{ K_n\left(m_sR\right)\frac{n}{R} - m_sK_{n+1}\left(m_sR\right)\right\}$$

$$\theta_{51} = \rho_f\omega^2 I_n\left(m_fR\right)$$

$$\theta_{52} = -\frac{\omega^2}{V_{p1}^{2}}\left[G + TC_1\right]K_n\left(m_{p1}R\right)$$

$$\theta_{53} = -\frac{\omega^2}{V_{p2}^{2}}\left[G + TC_2\right]K_n\left(m_{p2}R\right)$$

A8.6

Where the constants $N, F, Q, T$, and $C_i$ are the same as those defined in Equations 8.7 and 8.16, and $\chi = -\gamma_{12}/\gamma_{22}$.

In situations where the borehole wall is assumed to be impermeable, the coefficients $\theta_{ij}$ remain the same in rows 2, 3, and 4, while the first and last rows become

$$
\begin{pmatrix}
\theta_{11} \ \theta_{12} \ \theta_{13} \ \theta_{14} \ \theta_{15} \\
\theta_{21} \ \theta_{22} \ \theta_{23} \ \theta_{24} \ \theta_{25} \\
0 \ \ \theta_{32} \ \theta_{33} \ \theta_{34} \ \theta_{35} \\
0 \ \ \theta_{42} \ \theta_{43} \ \theta_{44} \ \theta_{45} \\
0 \ \ \theta_{52} \ \theta_{53} \ \theta_{54} \ \theta_{55}
\end{pmatrix}
\begin{pmatrix}
D_n \\
A_{1n} \\
A_{2n} \\
B_n \\
E_n
\end{pmatrix}
=
\begin{pmatrix}
S_{1n} \\
S_{2n} \\
0 \\
0 \\
0
\end{pmatrix}
$$

$$\theta_{11} = \frac{n}{R} I_n\!\left(m_f R\right) + m_f I_{n+1}\!\left(m_f R\right)$$

$$\theta_{12} = -\frac{n}{R} K_n\!\left(m_{p1} R\right) + m_{p1} K_{n+1}\!\left(m_{p1} R\right)$$

$$\theta_{13} = -\frac{n}{R} K_n\!\left(m_{p2} R\right) + m_{p2} K_{n+1}\!\left(m_{p2} R\right)$$

$$\theta_{14} = -\frac{n}{R} K\!\left(m_s R\right)$$

$$\theta_{15} = -i\kappa \left\{ K_n\!\left(m_s R\right) \frac{n}{R} - m_s K_{n+1}\!\left(m_s R\right) \right\} \qquad\qquad A8.7$$

$$\theta_{52} = \phi\!\left(C_1 - 1\right)\phi_{12}$$

$$\theta_{53} = \phi\!\left(C_2 - 1\right)\phi_{13}$$

$$\theta_{54} = \phi\!\left(\chi - 1\right)\phi_{14}$$

$$\theta_{55} = \phi\!\left(\chi - 1\right)\phi_{15}$$

# 9

# Qualitative and Quantitative Interpretation of Fracture Permeability by Means of Acoustic Full Waveform Logs

A number of geotechnical applications require the *in situ* characterization of naturally occurring and induced fractures. Some of the most important of these applications include high level radioactive waste disposal, stimulation of production from tight reservoirs, interpretation of the seismic properties of fault and shear zones, and estimation of the mechanical properties of fractured rock masses surrounding underground structures. Surface geophysical methods do not provide the spatial resolution required to resolve individual fractures or to interpret their properties. One practical approach to the *in situ* investigation of fractures located deep within a rock mass is by means of geophysical measurements in boreholes, including both conventional well logs and various new experimental techniques based on a number of new or expanded geophysical methods.

Acoustic full waveform logging has become one of the most important methods for fracture characterization in boreholes. Waveform logs provide useful information about the properties of fractures adjacent to the borehole because logging sources use frequencies that are high enough to provide spatial resolution at the scale of individual fractures, and low enough to penetrate into the formation beyond the annulus of rock affected by drilling. For example, logging frequencies in the range from 10 to 20 kHz correspond to wavelengths from 10 to 50 cm for seismic velocities ranging from 3.0 to 6.0 km/s. Information about the seismic properties of fractured rock also is important because seismic velocities and attenuation are closely related to the mechanical properties of rocks, and shear propagation is sensitive to the presence of fluids contained in fracture openings. Therefore, waveform log interpretations are useful in applications related to the mechanical properties or hydraulic conductivity of fractured rock masses. As a result, acoustic full waveform logging has become an important means for the investigation of fractured rocks, including the interpretation of fracture permeability *in situ*.

## 9.1    WAVEFORM LOGS IN FRACTURED ROCKS

The earliest applications of acoustic waveform logs to the interpretation of fracture properties were made in the detection of fractures on so-called "three dimensional" acoustic logs. In this form of logging, the entire pressure signal received at one or more acoustic receivers is displayed in the standard seismic format where positive pressure fluctuations are printed solid black and negative fluctuations are left blank. In unfractured rock, this display represents the waveforms as a series of parallel wavefronts. The wavefronts become less coherent over depth intervals where seismic velocities vary. Fractures intersecting the borehole produce large discontinuities for wave energy propagation along the borehole, resulting in distinctive reflection patterns ("chevrons") on the wavefront display (Christiansen, 1964). Other discontinuities such as bed contacts might produce similar reflections in the log, but it was assumed that the larger amplitude of fracture reflections and the lithologic context would allow correct interpretation.

One of the most significant difficulties in the application of "three-dimensional" acoustic logs to fracture interpretation was the inability of the method to distinguish near-vertical fractures. Discontinuities aligned along the borehole would not be orientated so as to interfere with wave propagation along the borehole wall, and would not be indicated by reflections in the data. The early theory on deep reservoir fracturing indicated that minimum principal stresses would almost always be horizontal, so that open fractures in deep petroleum reservoirs would be vertical. For this reason, a circumferential acoustic logging system (Vogel and Herolz, 1977; Guy, 1987) was designed to measure the amplitude of ultrasonic signals (50 to 100 kHz) propagating between receivers aligned along the circumference of the borehole. This device is commerically available from at least one major service company, but results are not simple to interpret, and additional equipment development appears necessary before this potentially useful technique can achieve its full potential.

Interest in fracture interpretation problems other than deep petroleum reservoir production from fractures renewed interest in using acoustic full waveform logs as a qualitative means for investigating fractures (Witherspoon et al., 1981). Acoustic waveforms plotted at successive depths and adjacent to fractures intersecting a borehole are illustrated in Figure 9.1 for isolated fractures aligned nearly perpendicular and parallel to the borehole axis. The waveforms indicate that seismic wave propagation is affected by both of the fracture sets in Figure 9.1. Waveforms representing transmission through unfractured rock above and below the isolated fractures are almost identical, confirming that distortion and disruption of the characteristic wave signature are associated with transmission across the fractures.

The degree of wave disruption and decrease in average wave amplitude in waveform data such as that in Figure 9.1 appear to be related to the size of the fracture. This general association was observed by using the borehole televiewer

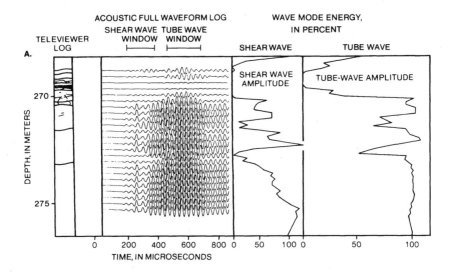

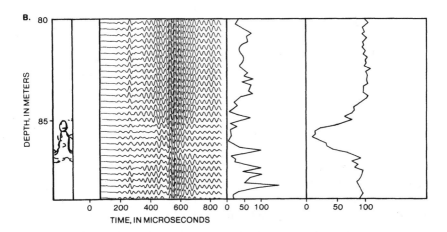

FIGURE 9.1.    Acoustic full waveform log data and amplitude logs calculated from that data compared to televiewer log for (A) interval containing nearly horizontal fractures and (B) interval containing nearly vertical fracture.

to provide independent information on the location and relative size of the fracture. The televiewer is an ultrasonic wall scanning device operating at frequencies near 1 MHz and using the intensity of reflection of acoustic energy off of the borehole wall to provide a photographic image of fractures (Zemanek et al., 1969, 1970; Paillet et al., 1990). The borehole-fracture intersection scatters acoustic energy such that the location of the fracture is indicated by a dark line

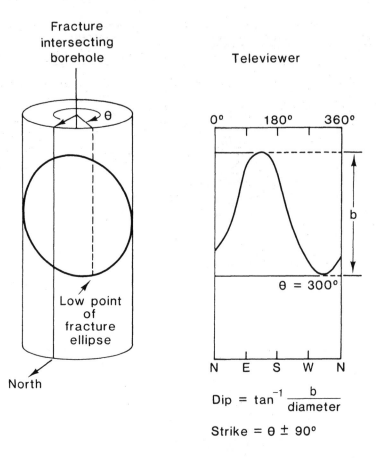

FIGURE 9.2. Interpretation of fracture strike and dip using televiewer log data.

on the televiewer log display. The vertical interval over which the fracture intersects the borehole is used to determine the orientation of the fracture, while the relative width of the fracture image on the televiewer log is taken as a qualitative indication of fracture aperture (Figure 9.2). One important limitation in use of the televiewer is the very shallow depth of investigation provided by televiewer logs. The combination of televiewer and acoustic waveform logs appeared to be an ideal means by which to approach fracture characterization. The televiewer log could be used to identify fracture locations and fracture orientation, and the waveform log interpretation used to infer the properties of the fracture in the region beyond the thin annulus disturbed by drilling. Such combined plots of televiewer logs and acoustic waveforms, or "acoustic signature logs", are still of great use today in assessing the quality of rocks for various engineering applications.

The qualitative association between waveform log amplitude attenuation and fractures indicated that quantitative relations between fracture properties such as aperture, orientation, and hydraulic conductivity might be established. One of the earliest attempts related the amplitude of successive compressional arrivals according to the formula (Lebreton et al., 1978)

$$ I_c = \frac{E_2 + E_3}{E_1} \qquad\qquad 9.1 $$

where $I_c$ is a compressional wave amplitude index, and $E_1$, $E_2$, and $E_3$ are the maximum amplitudes of the first three half-cycles of the compressional wave. These authors indicate that there is a statistically significant correlation between fracture permeability and $I_c$. Although this application appeared to be successful in the cases presented by Lebreton et al. (1978), few other investigations have reproduced these results.

Paillet (1980) attempted to relate amplitude of shear arrivals to permeability under the assumption that shear propagation would be more sensitive to the presence of fluids in permeable fracture openings than compressional waves. Initial investigation indicated that shear amplitude measured by centering the amplitude window on expected shear arrivals did correlate with independent, semiquantitative measurements of fracture aperture. However, the correlation applied only to fractures that were transverse to the borehole axis, and required previous knowledge of shear velocity in order to specify time windows for amplitude measurements. At the same time, Paillet (1980) and Paillet and White (1982) noted large shear amplitude increases associated with the edges of fracture anomalies (Figure 9.3). These were attributed to mode conversion in which the interference in seismic transmission across fractures resulted in enhanced acoustic energy transmission when the acoustic receiver was opposite the fracture opening.

Although Paillet (1980) found significant complications in the interpretation of shear arrival amplitudes, amplitude windows arranged to coincide with wave energy propagating at acoustic velocity in the borehole fluid were found to correlate with measured permeabilities; examples are illustrated in Figures 9.1 and 9.3. Earlier calculations by Rosenbaum (1974) indicated that these late-arriving wave trains were strongly attenuated by borehole wall permeability in models for porous rocks based on the Biot (1956a,b) formulations. Paillet (1980) inferred that the viscous dissipation mechanism believed to cause the attenuation in the Rosenbaum (1974) model might explain the fracture attenuation noted in his data. Subsequent analysis of waveform data and theoretical models of waveform propagation across fractures have confirmed the relationship between qualitative estimate of permeability and the attenuation of trapped wave modes traveling along the borehole at approximately acoustic velocity of the borehole fluid. However, the attenuation and permeability correlation applies primarily to only one of a series of possible modes. The emphasis on this single mode and its properties in fracture interpretation introduces one important consideration: tube waves must be excited in the borehole. The mode excitations discussed in

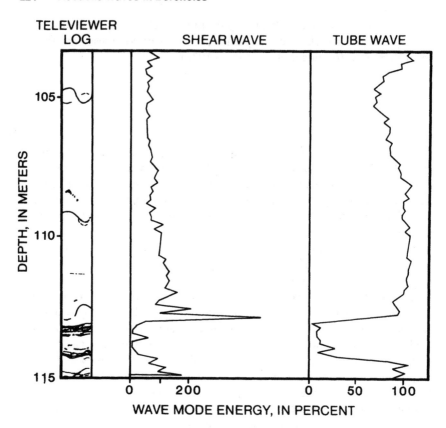

FIGURE 9.3.    Wave mode amplitude calculated for shear head wave and tube wave modes compared to televiewer log for an interval where shear mode energy is enhanced by mode conversion.

Chapters 3, 4, and 5 indicate an upper frequency limit beyond which tube waves will not be excited with the required amplitude. That upper frequency limit depends upon both the borehole diameter and the seismic velocities of the formation.

## 9.2    DIRECT CORRELATION BETWEEN TUBE WAVE ATTENUATION AND FRACTURE PERMEABILITY

Although it has proven difficult to develop a completely acceptable theoretical relationship between fracture permeability and tube wave attenuation (Paillet et al., 1989), direct empirical correlations between independent measurements can be established. One of the major obstacles to the direct empirical calibration of tube wave attenuation in this way is the difficulty encountered in measuring permeability. In sedimentary formations, permeability may be measured using recovered core samples (Burns et al., 1988; Burns, 1989). However, these

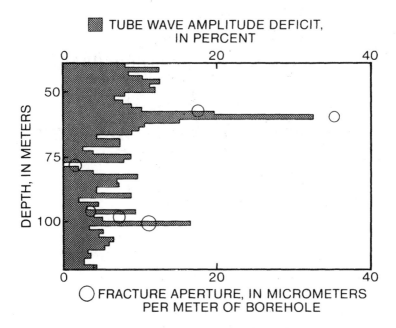

FIGURE 9.4.    Comparison of tube wave amplitude deficit integrated over 3-m intervals with equivalent fracture aperture determined in packer tests showing correlation of amplitude deficits with permeability; Chalk River, Ontario (Paillet, 1983a).

samples are smaller than the volume of rock over which tube waves average rock properties adjacent to boreholes. Useful correlations between tube wave attenuation in such formations would require the application of an averaging filter to a large number of such permeability measurements. Permeability measurements in fractured crystalline rocks are even more difficult to obtain. Very few fractures can be recovered intact in core samples, and these likely provide too small a sample of the fracture passage to be representative of fracture hydraulics. The accepted method for measuring fracture permeability *in situ* is by means of straddle packer isolation and injection tests (Zeigler, 1976; Hsieh et al., 1983; Pickens et al., 1987). These measurements give the integrated permeability or transmissivity of a vertical interval of borehole containing one or more fractures. In studies where a large number of straddle packer tests have been made, the resulting permeability distribution can be used to calibrate tube wave attenuation measurements.

   Paillet (1983a) gave a representative example of empirical correlation of tube wave attenuation and measured fracture permeability (Figure 9.4). A series of prolonged straddle packer tests were made in fractured metamorphic rocks at a site on the Canadian Shield near Chalk River, Ontario. Calculations demonstrated that the 35 kHz acoustic logging tool generated waveforms in a 7.5 cm diameter borehole without contribution from pseudo-Rayleigh waves superimposed on the tube wave (Figure 9.5). Tube wave amplitude was calculated from

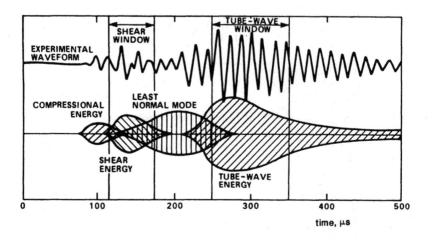

FIGURE 9.5.    Waveform log data in unfractured granitic rocks compared to mode partition predicted from a theoretical model. The model illustrates time windows used to construct shear and tube wave amplitude logs; Chalk River, Ontario (Paillet, 1983a).

the waveforms using the time window shown in the figure. Wave mode energy (the quantitative measure of tube wave amplitude used throughout this analysis) was calculated by summing the square of the departure from baseline for each point in the time window, and multiplication by a half-cosine filter in order to minimize the effects of wave peaks at the filter edges. In addition, tube wave amplitude (expressed as the average wave energy in intervals of unfractured rock) was found to increase with depth. This difference in amplitude of excitation is illustrated by the difference in background tube wave amplitudes between Figures 9.1A and 9.1B. This was attributed to the effect of hydrostatic pressure on the efficiency of source excitation.

One important difficulty in comparing tube wave attenuation with measured permeabilities is the possibility of straddle packer settings spanning a number of fracture sets in a single measurement. In Figure 9.4, the diameter of the circle indicating packer permeability data corresponds to the length of the isolated interval. The corresponding interval of borehole associated with each tube wave amplitude measurement is the source to receiver spacing on the logging tool (about 60 cm in Figure 9.4). The tube wave amplitude data was made approximately equivalent to the straddle packer data by combining tube wave amplitude anomalies over each 2-m interval. Tube wave amplitude decreases were expressed as the percent of average amplitude in adjacent unfractured intervals at each fracture anomaly in the tube wave amplitude data. The percent difference between amplitude in fractured zones and background amplitude in adjacent unfractured rock defines the tube wave amplitude deficit. The amplitude deficit for each 2-m interval was taken as the sum of all the deficits in the interval. This is equivalent to summing the transmissivities of fracture sets in each interval under the assumption that individual tube wave amplitude deficits are proportional to the transmissivity of individual fractures. The data in Figure 9.4 indicate that these interval-summed amplitude deficits can be correlated with the permeability from packer tests, and the correlation used to provide an empirical calibration of tube wave amplitude data.

The results illustrated in Figure 9.4 are typical of results obtained at a number of Canadian Shield sites (Paillet, 1988; Paillet and Hess, 1986, 1987). Consistent results were obtained for isolated sets of fractures, but the tube wave amplitude method appeared to overestimate fracture permeability in extensively fractured intervals. These inconsistencies between tube wave data and packer tests appeared to reflect the combined effects of fracture opening adjacent to the borehole, and plugging of fractures further into the formation by the mobilization of gouge clays in weathered fault zones. Inspection of borehole walls with remote television cameras supported these conclusions in that fracture fillings were seen to have been eroded, and permeabilities measured by straddle packer tests appeared orders of magnitude less than would be inferred from the number and size of fracture openings adjacent to the borehole.

One of the most interesting and useful conclusions to be drawn from experimental studies of tube wave attenuation is the apparent insensitivity to fracture orientation. Paillet and Hess (1987) found that isolated, nearly vertical fractures produced tube wave anomalies similar to those of fractures transverse to the borehole axis, except that the amplitude anomaly was stretched out along the vertical length of the fracture-borehole intersection. To a first approximation, the tube wave energy attenuation correlated with the transmissivity between the acoustic source and receiver. This conclusion is supported by the borehole model calculations of Tang and Cheng (1989) and Poeter (1987). In both models, tube wave amplitude decreases almost linearly with source to receiver separations (and hence length of fracture intersection) in scale models of boreholes with planar openings of a constant aperture representing vertical fractures. The total transmissivity of the opening between source and receiver also increases linearly with source to receiver separation, in support of the theory relating tube wave attenuation to fracture transmissivity.

All of these data indicate that empirical calibration and interpretation of tube wave amplitude logs can be an effective semiquantitative means for estimating fracture permeability in boreholes where there is additional supportive data in the form of borehole televiewer, remote television, or core fracture description. The measured energy attenuation of tube waves appears to correlate with the transmissivity or vertically integrated hydraulic conductivity of the set of fractures intersecting the borehole between source and receiver.

## 9.3   TUBE WAVE INTERPRETATION MODELS

In many geotechnical applications, time limitations and expense rule out packer permeability tests. At the same time, specific predictive models relating tube wave attenuation to fracture permeability would lend considerable confidence in the interpretation. Numerous authors have proposed fracture models, but none of those presented so far in the literature appear to define a relationship between fracture permeability and tube wave attenuation that agrees in all ways with observations.

The effects of fracture permeability on tube wave amplitude can be approximated by calculating tube wave propagation in a borehole intersected by a plane, parallel fracture filled with a viscid fluid. The dynamic effects of a tube wave propagating along a borehole intersected by such a borehole have been calculated under the assumption that fracture aperture is much larger than the viscous boundary larger for oscillatory motion in the fracture by Hornby et al. (1989). The results of calculations using such models qualitatively agree with field experiments at relatively low frequencies (1 to 3 kHz) but are quite different from results obtained at typical logging frequencies (10 to 30 kHz) used in mining and hydrologic applications. In particular, the models yield results that are dominated by tube wave reflections. Extensive studies of waveform log anomalies associated with isolated fractures indicate that coherent tube wave reflections are not encountered using acoustic source frequencies greater than 10 kHz, and are only rarely documented using a 5 kHz source (Paillet 1984, 1988) (Figure 9.6).

Finite difference models of acoustic wave propagation along boreholes intersected by permeable fractures yield results very similar to the small boundary layer fracture models (Stephen et al., 1985; Bhashvanija, 1983). The waveforms calculated by the finite element method are dominated by tube wave reflections, and fracture apertures are required to be as large as several centimeters to produce a measurable effect on tube wave amplitudes.

The most complete model for tube wave propagation across permeable fractures is given by Tang and Cheng (1989). Their analysis demonstrates that the equations of motion governing oscillatory flow of frequency $\omega$ in a thin, planar opening can be expressed as the solution of two partial differential equations:

$$\nabla^2 \phi + \frac{\omega^2}{V_f^2 - \frac{4}{3} i\omega \frac{\eta}{\rho}} = 0$$

$$\nabla^2 \psi + \frac{i\rho\omega}{\eta} \psi = 0$$

9.2

where the fluid velocity, $V$ is expressed in terms of an acoustic wave potential, $\phi$, and a viscous shear potential, $\zeta$:

$$V = \nabla\phi + \nabla \times \zeta$$

$$\zeta = (0, \psi, 0)$$

9.3

Expressing these equations in cylindrical coordinates, and enforcing nonslip boundary conditions at the top and bottom of the fracture opening, results in a characteristic equation giving those combinations of frequency and wave number for which there can be wave propagation within the fracture:

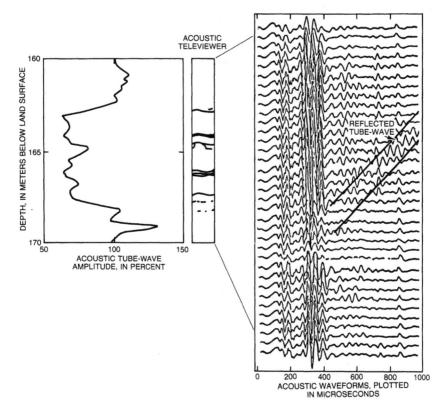

**FIGURE 9.6.**    Acoustic waveform log data obtained in granite with experimental 5 kHz source illustrating weak tube wave reflections from a large, subhorizontal fracture (Paillet, 1988).

$$\kappa^{2}\tan\left(\alpha_{2}\,\frac{a}{2}\right)+\alpha_{1}\alpha_{2}\tan\left(\alpha_{1}\,\frac{a}{2}\right) \qquad\qquad 9.4$$

$$\alpha_{1}^{2}=\frac{w^{2}}{v_{f}^{2}-\frac{4}{3}iwv}-k^{2}$$

$$\alpha_{2}^{2}=\frac{iw}{v}-k^{2}$$

where $\kappa$ is the wave number, $v$ is the kinematic viscosity of the fluid ($\eta/\rho$), and $a$ the half-width of the planar fracture opening. The variation of wave velocity with fracture aperture is illustrated in Figure 9.7A. The acoustic wave velocity is seen to decrease rapidly when the fracture aperture becomes smaller than the viscous boundary layer thickness given by

$$\delta=\sqrt{\frac{2\eta}{\rho_{f}\omega}} \qquad\qquad 9.5$$

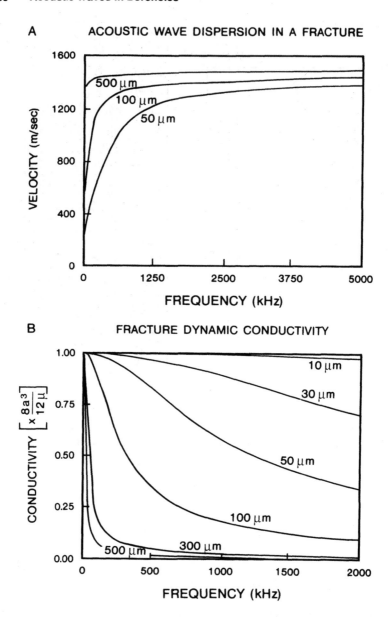

FIGURE 9.7.    Properties oscillatory flow in planar fractures: (A) velocity of propagating waves for various apertures and (B) dynamic hydraulic conductivity expressed as the fracture of conductivity for steady flow.

If the flow in the fracture is driven by an oscillatory pressure source applied at a cylindrical opening of radius $R$, the resulting flow in the fracture is shown to have the form

$$q = -CP'$$ 9.6

where $q$ is the average flow and $P'$ is the radial pressure gradient in the fracture at the surface of the borehole where the pressure is applied; and the dynamic hydraulic conductivity of the fracture is given by

$$C(\omega) = \frac{i\omega a}{\kappa^2 V_f^2 \rho_f}$$ 9.7

In the limit of very low frequencies, Equation 9.4 yields

$$\kappa^2 = \frac{12i\omega\eta}{\rho_f V_f^2 a^2}$$ 9.8

which gives a hydraulic conductivity in agreement with the cubic law for fracture flow (Witherspoon et al., 1981) when substituted in Equation 9.6:

$$\lim_{\omega \to 0} C(\omega) = \frac{a^3}{12\eta}$$ 9.9

Dynamic conductivities are given as a function of frequency for various fracture apertures in Figure 9.7B. If the pressure produced by a tube wave propagating along the fracture is expressed as

$$P_0 = E_I I_0\left(m_f r\right) e^{i(\kappa z - \omega t)}$$

$$m_f^2 = \kappa^2 - \frac{\omega^2}{V_f^2}$$ 9.10

Tang and Cheng (1989) showed that the reflected and transmitted tube wave energy is given by

$$\frac{E_R}{E_I} = -\frac{Y}{Y+1} \qquad\qquad \frac{E_T}{E_I} = -\frac{1}{Y+1}$$

$$Y = \frac{\rho_f \omega}{2\kappa} C\left\{\frac{\kappa m_f I_0\left(m_f R\right) H_1^{(1)}\left(m_f R\right)}{I_1\left(m_f R\right) H_0^{(1)}\left(m_f R\right)}\right\}$$ 9.11

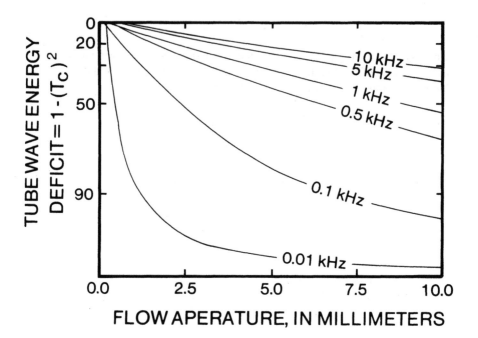

FIGURE 9.8.    Theoretical transmitted tube wave amplitude and associated tube wave energy deficit calculated a function of fracture aperture and tube wave frequency using the plane-fracture model of Tang and Cheng (1989); calculations based on water, granite, and a borehole diameter of 7.5 cm.

where $C$ is the dynamic conductivity defined in Equation 9.7. Note that the sum of transmitted and reflected wave energy does not necessarily equal the incident energy because an additional amount of wave energy is propagated outward into the fluid filling the fracture. This solution is restricted in its application to large values of fracture aperture, $2a$, because derivation of Equation 9.11 requires that $a$ is much less than the wavelength of the tube wave driving the flow.

The transmitted tube wave amplitude and energy is plotted as a function of frequency in Figure 9.8. The transmitted tube wave energy is the quantity that is used to plot the tube wave amplitude logs given in Figures 9.1 and 9.3. However, the results in Figure 9.8 indicate that fracture apertures greater than several centimeters are required to produce significant tube wave attenuation at frequencies greater than 5 kHz. The theory also indicates that measurable tube wave reflections are generated and that there is substantial frequency dependence in the transmission coefficients. All of these effects are observed in measurements made using laboratory scale models of fractures where artificial openings of constant aperture intersect boreholes drilled in solid blocks of metal or plastic. The magnitude of the measured transmission coefficients also agrees with the theory for these fracture models (Figure 9.9).

Although there is consistent agreement between predicted tube wave transmission coefficients and theoretical calculations for laboratory models, the

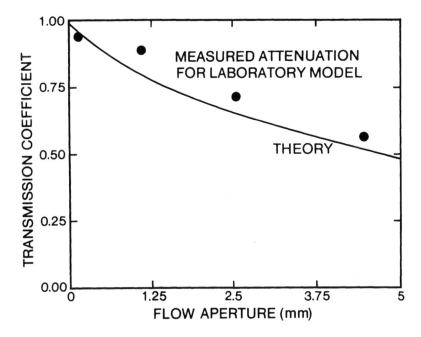

FIGURE 9.9.    Comparison of transmission coefficient predicted by plane fracture theory with transmission coefficient determined from measured attenuation using a laboratory model; 100 kHz in 0.5 cm borehole in aluminum.

characteristics of tube wave measurements in the boreholes penetrating natural fracture zones are different from those predicted by the theory. Independent measurements of fracture aperture indicate that tube wave attenuation greater than 50% can be produced by fractures with apertures as small as 0.5 mm (Paillet, 1983a). Furthermore, there is no apparent frequency dependence (as long as frequencies are greater than a few kilohertz) in studies where more than one tube wave frequency is used, and coherent tube wave reflections are rarely measured (Paillet, 1984). Paillet et al. (1989) proposed a model in which natural fractures are composed of a network of flow tubes between asperities on fracture faces. The model indicates that flow tubes as small as 1 cm in diameter can produce measurable tube wave attenuation, and yet would conduct flow during steady injection tests equivalent to an infinite plane fracture with an aperture as small as 1 mm, depending upon the estimate used for flow tube spacing. The flow tube model for fractures also removes frequency dependence from the tube wave response and allows for degradation of tube wave reflections by scattering. However, the flow tube model still seems to require equivalent single fracture apertures greater than a few millimeters in order to produce measurable tube wave attenuation.

One predictive model which has been used to relate tube wave attenuation to effective hydraulic aperture of fractures ignores wave dynamics (Mathieu, 1984; Algan and Toksöz, 1986). Instead, the observation of little or no reflected wave

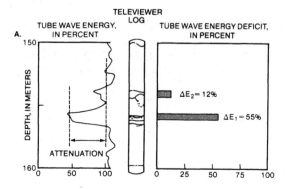

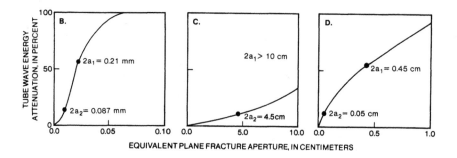

FIGURE 9.10.   Example of fracture aperture interpretation using theoretical models: (A) measurement of tube wave energy deficit, and interpretation of fracture aperture using the (B) Mathieu (1984), (C) Tang and Cheng (1989), and (D) Paillet et al. (1989) models for 34 kHz acoustic source in a 7.5 cm diameter borehole.

energy is used to construct an approximate energy balance. In this model, all tube wave energy decreases are attributed to viscous dissipation in the fracture opening. Although this model has been criticized for its lack of basis in theory (Norris, 1989), one may argue that the results give the effective size of a fracture opening required to produce an equivalent decrease in the oscillatory motion of the tube wave. Under the assumption that neither the tube wave attenuation model nor the single plane fracture model used in the interpretation of packer tests represent real fractures, the energy balance model of Mathieu (1984) may provide a useful scale for the effects of fracture permeability on tube waves. Results of the Mathieu (1984), Tang and Cheng (1989), and Paillet et al. (1989) theoretical models are illustrated for typical tube wave amplitude data in Figure 9.10. Fracture apertures estimated from the size of crystals deposited in openings in fractures identified on core range from 0.1 to 1.00 mm for the fractures indicated in Figure 9.10A.

Models relating tube wave attenuation to fracture permeability provide additional information of use in the interpretation of acoustic full waveform logs in fractured rocks beyond the direct calibration of fracture permeability. Even though Mathieu (1984) did not provide a correct dynamic model for the propagation of tube waves along boreholes intersected by fractures, the theory provides a lower limit on the effective hydraulic aperture of the fracture (Norris, 1989). The higher order viscous terms in the model provide the most effective decrease in wave energy at small fracture apertures, so the assumption that all attenuation is attributed to the viscous term gives the smallest aperture that could produce the observed tubewave attenuation. At the same time, fracture model calculations indicate the distance away from the borehole over which the tubewave attenuation responds to fracture permeability. The results of Algan and Toksöz (1986) demonstrate that tube wave attenuation is influenced by permeability over a distance corresponding to one to several wavelengths (1 to 5 at acoustic frequencies near 10 kHz), with the larger depths of penetration corresponding to the more permeable fractures.

In summary, a completely satisfactory theoretical model for tube wave attenuation and reflection in boreholes does not exist. The plane, infinite, uniform aperture fracture model seems to apply to tube wave interaction with fractures at frequencies less than a few kHz in typical boreholes. This model can be used to relate the amplitude of the reflections to fracture apertures using the theory given by Hornby et al. (1989). However, this interpretation method is not as directly related to fracture permeability as the hypothesized direct relationship between tube wave attenuation and fracture transmissivity at frequencies greater than 5 kHz. The plane fracture models do not appear to apply to this frequency range, and the frequencies at which the transition between the low-frequency regime and the high-frequency attenuation regime occurs may depend upon the details of the distribution of asperities and flow tubes within individual fractures. In spite of these theoretical difficulties, the consistent agreement between the fracture apertures estimated from the assumption that all attenuation is attributed to simple viscous dissipation and the results of packer isolation tests indicates that there is a relationship between tube wave attenuation and transmissivity at frequencies greater than 5 kHz in typical boreholes.

## 9.4    FIELD STUDIES — EVALUATION OF TUBE WAVE ATTENUATION MODELS AND COMPARISON WITH OTHER ESTIMATES OF FRACTURE PERMEABILITY

The consistent association of tube wave attenuation with permeability in both porous sediments and fractured crystalline rocks appears well established in the literature. Qualitative interpretation of fracture permeability on the basis of tube wave interpretation has become an effective means for fracture characterization in several ongoing studies (Paillet and Hess, 1987, 1988). However, quantitative evaluation of the method is made very difficult by problems associated with obtaining reliable permeability information for comparison. The best method for verification of tube wave estimates of fracture permeability is packer isolation and injection tests (Hsieh et al., 1983; Pickens et al., 1987). Extensive fracture permeability studies have been performed at two crystalline rock sites: Mirror

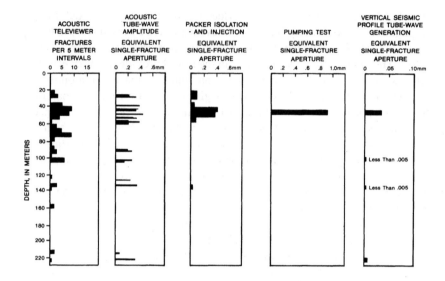

FIGURE 9.11. Comparison of fracture permeability distributions in 15 cm diameter borehole at Mirror Lake estimated using various interpretation methods (from Paillet et al., 1987).

Lake, New Hampshire (Paillet et al., 1987; Hardin et al., 1987), and the Atomic Energy of Canada Limited (AECL) crystalline rock research site near Lac duBonnet, Manitoba (Green and Mair, 1983; Davison et al., 1982; Katsube and Hume, 1987; Paillet, 1989, 1991). Fracture permeability measurements at these two sites provide a quantitative test of the tube wave method in the estimation of fracture permeability.

Fracture identification and characterization was carried out in a series of boreholes at the Mirror Lake site using a broad range of geophysical techniques. Bedrock in the Mirror Lake area consists of micaceous schist intruded by extensive bodies of quartz monzonite and other rocks (Paillet et al., 1987). Core was not obtained, but lithology descriptions from cuttings were provided for each of the boreholes. All boreholes were cased through approximately 15 m of glacial drift, were 16 cm in diameter, and varied from 100 to more than 200 m in depth. Acoustic waveform and televiewer logs were obtained in the boreholes, in addition to a full suite of conventional geophysical well logs. Acoustic waveforms were obtained using a relatively low frequency (12 kHz) source to insure that tube waves could be identified in the recorded waveforms. Subsequent models verified that tube waves were generated, although somewhat contaminated by the Airy phase of the first pseudo-Rayleigh mode. Because of the the possibility that the pseudo-Rayleigh wave might have affected interpretations, a limited number of boreholes also were logged with an experimental low-frequency logging system operating at 5 kHz, reproducing the tube wave calculations made using the 12 kHz source.

The results of the tube wave interpretations of fracture permeability are summarized in Figure 9.11. The various representations of fracture permeability are given in order of increasing scale of investigation from left to right in the figure. The smallest scale results are represented in terms of the number of possibly open fractures in each 5-m interval of borehole interpreted from the televiewer log. This fracture distribution is compared to the acoustic tube wave interpretation, where the effective hydraulic aperture of each apparently permeable fracture is given in millimeters as calculated from the Mathieu (1984) method. These results may then be compared to the permeability given by packer tests, hole-to-hole pumping tests, and surface to borehole vertical seismic profile (VSP) interpretation. The packer tests agree rather well with the tube wave estimations of permeability if the individual fracture permeabilities are combined by summing the transmissivities (proportional to the cube of effective aperture; Snow, 1965) within the intervals used for the packer tests. The same order of magnitude was given for fracture permeability in the main fracture zone connecting the network of boreholes during the hole-to-hole pumping tests.

The small differences in the actual values given for the permeabilities in Table 9.1 could be attributed to the differences in scale of investigation associated with each method of permeability estimation. However, the VSP method does not appear to agree even approximately with the tube wave interpretation. In this method, the amplitude of low-frequency tube waves generated in the borehole by the passage of seismic waves across the fracture mouth are related to fracture permeability (Huang and Hunter, 1981; Hardin and Toksöz, 1985). This discrepancy appeared even more serious at first, because the VSP interpretation by Hardin et al. (1987) indicated a nearly horizontal fracture zone, whereas televiewer logs indicated most fractures were steeply dipping. The results were clarified by the hole-to-hole pumping tests, in which the interconnection between boreholes was inferred to be an irregular path composed of individual segments of dipping fractures (Paillet et al., 1987). The reasonable agreement between results from the tube wave interpretation, packer tests, and hole-to-hole pumping otherwise appeared to support the application of the tube wave method (with the interpretation method of Mathieu, 1984) and to suggest a possible oversimplification in the interpretation of the VSP method.

The AECL study site in Manitoba provided approximately the same set of data for comparison with tube wave interpretations, along with extensive fracture descriptions from core. Tube waves were obtained using a 34 kHz source similar to that described for the work at Chalk River, Ontario (Paillet, 1983a). Fracture permeabilities distribution in a 1200-m deep core hole interpreted from tube wave attenuation using the Mathieu (1984) method are compared to the distribution of fractures identified on core samples in Figure 9.12. The results of the tube wave interpretation at the AECL site are compared to other data in Figure 9.13 for a core hole where packer tests were performed to produce an independent measurement of the permeability distribution. The VSP data were not interpreted by the method used at Mirror Lake, but are given as a percent of the largest fracture response at 270 m in depth in Figure 9.13. This response is approximately proportional to fracture transmissivity (Hardin and Toksöz, 1985; Hardin et al., 1987). The relative proportion of the permeability response

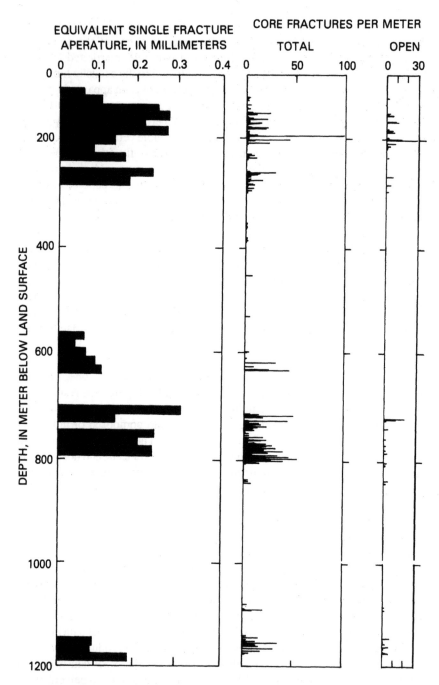

FIGURE 9.12.  Comparison of fracture permeability distribution estimated using the Mathieu (1984) model with distribution of permeable fractures identified by core logger; 7.5 cm diameter borehole in granite (from Paillet, 1988).

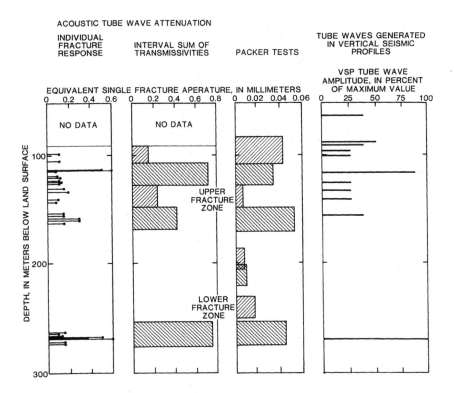

FIGURE 9.13. Comparison of fracture permeability distributions in a 7.5 cm diameter borehole at the AECL Manitoba site estimated using various interpretation methods.

on the VSP data agrees with the tube wave interpretation. The data in Figure 9.13 indicate the same sort of qualitative agreement between the tube wave interpretation and the packer test results. However, the packer tests indicate nearly a full order of magnitude lower values for effective hydraulic aperture. Because fracture transmissivity is proportional to the third power of aperture, this is a substantial difference. The discrepancy is even more difficult to explain because the Mathieu (1984) method used to interpret the data is known to predict the minimum fracture aperture required to produce the observed tube wave attenuation.

The quantitative difference between permeabilities estimated using the tube wave attenuation method and the permeabilities measured by the packer isolation and injection technique appear difficult to explain in comparison to the close agreement found by Paillet et al. (1987) at Mirror Lake, New Hampshire. The difference in acoustic frequencies used was originally questioned, but repeat logging of the AECL core holes in Manitoba with an experimental low-frequency source did not affect the fracture interpretation (Figure 9.14). The scale of investigation of the two methods may be important because the packer

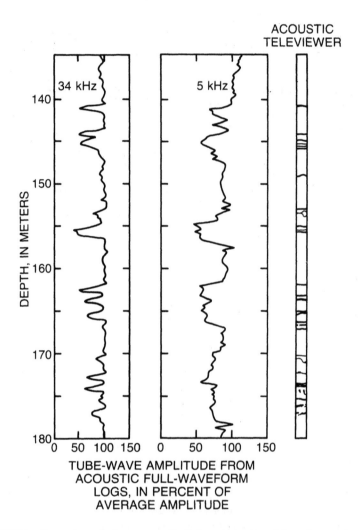

FIGURE 9.14.   Comparison of tube wave amplitude logs obtained in a 7.5 cm diameter borehole at the AECL Manitoba site using 34 kHz and five sources (from Paillet, 1988).

tests involve a much larger radial depth of investigation than the tube wave measurements. Deposits of clay minerals generated by rock weathering were observed to line some fractures in core samples. These weathering products may have been eroded from the fractures adjacent to the borehole during drilling, accounting for the larger permeabilities inferred from the tube wave analysis. However, the difference between the packer test results and the tube wave interpretation method may relate to the interpretation model based on a planar, fluid-filled fracture opening. Natural fractures may more closely be modeled as a set of irregular flow tubes located between asperities in contact along the fracture plane. Recent results indicate that the effects of high-frequency oscillation on hydraulic transmissivity (defined as the ratio of discharge to pressure gradient in such irregular flow tubes) could explain the differences between permeability interpretations based on ultrasonic tube wave oscillations and steady injection tests (Tang and Cheng, 1988; Paillet et al., 1989).

As of the publication of this monograph, one very recent model does appear to account for the relatively large tube wave attenuation (small transmission coefficients) associated with the small-aperture fractures indicated in Figure 9.13, and for the relatively weak frequency dependence indicated in Figure 9.14. This model approximates the fracture as a thin layer of porous material of given porosity, permeability, and thickness. An example is given in Figure 9.15, where model results are compared to the measured amplitudes of reflected and transmitted tube waves (Tang et al., 1991). Although the model is complex enough to require assumptions about the mechanical properties of the rock in the fracture zone in addition to estimates of fracture zone orientation and thickness inferred from the televiewer log, there is close agreement between the measured amplitudes of transmitted and reflected waves and those predicted by the model. These results demonstrate that this class of fracture model can account for the relatively large attenuation and weak reflections that characterize tube wave transmission across natural fractures at the Manitoba study site. The model also indicates that the effects of oscillatory flow induced in the fracture by the tube wave need to be considered when comparing the amplitude of tube wave attenuation or reflection with the estimates of fracture zone permeability obtained from quasi-steady flow during straddle packer isolation and injection tests (Tang and Cheng, 1988).

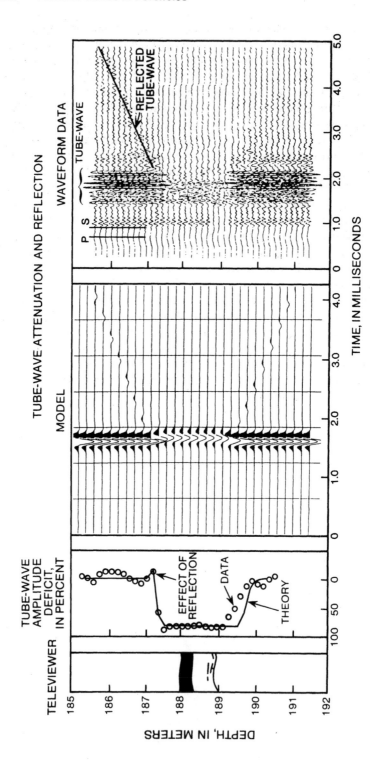

FIGURE 9.15.   Theoretical model of tube wave reflection and attenuation in propagation along a borehole intersected by a thin permeable layer compared to waveform data indicating the ability of the theory of Tang et al. (1991) to predict relative attenuation and intensity of reflections for 5 kHz tube waves in a 15–cm–diameter borehole in granite.

# References

**Abramowitz, M. and Stegun, I. A., Eds.,** Handbook of Mathematical Functions, National Bureau of Standards, U.S. Government Printing Office, Washington, D.C., AMS 55, 1046, 1964.

**Aki, K., Fehler, M., Aamodt, R. L., Albright, J. N., Potter, R. M., Pearson, C. M., and Tester, J. W.,** Interpretation of seismic data from hydraulic fracturing experiments at Fenton Hill, New Mexico, hot dry rock geothermal site, *J. Geophys. Res.,* 87, 936, 1982.

**Aki, K. and Richards, P. G.,** *Quantitative Seismology, Theory and Methods,* W.H. Freeman and Co., San Francisco, 932, 1980.

**Algan, U. and Toksöz, M. N.,** Depth of fluid penetration into a porous permeable formation due to sinusoidal pressure source in a borehole, *Log Anal.,* 27, 30, 1986.

**Alterman, Z. S. and Karal, F. C.,** Propagation of elastic waves in layered media by finite difference methods, *Bull. Seismol. Soc. Am.,* 58, 367, 1968.

**Alterman, Z. S. and Loewenthal, D.,** Computer generated seismograms, in *Methods in Computational Physics,* Vol. 12, Bolt, B. A., Ed., Academic Press, Inc., New York, 35, 1972.

**Anderson, D. L. and Archambeau, C. B.,** The anelasticity of the earth, *J. Geophys. Res.,* 69, 2071, 1964.

**Auriault, J. L.,** Dynamic behaviour of a porous medium saturated by a newtonian fluid, *Int. J. Eng. Sci.,* 18, 775, 1980.

**Auriault, J. L., Borne, L., and Chambon, R.,** Dynamics of porous saturated media, checking of the generalized law of Darcy, *J. Acoust. Soc. Am.,* 77, 1641, 1985.

**Azimi, Sh.A., Kalinin, A. V., Kalinin, V. V., and Pivovarov, B. L.,** Impulse and transient characteristics of media with linear and quadratic absorption laws, *Izv. Physics of the Solid Earth,* Feb., 88, 1968.

**Baker, L. J.,** The effect of the invaded zone on the full wavetrain acoustic logging, *Geophysics,* 49, 796, 1984.

**Baker, L. J. and Winbow, G. A.,** Multipole P-wave logging in formations altered by drilling, *Geophysics,* 53, 1207, 1988.

**Banthia, B. S., King, M S., and Fatt, I.,** Ultrasonic shear-wave velocities in rocks subjected to simulated overburden pressure and internal pore pressure, *Geophysics,* 30, 117, 1965.

**Bear, J.,** *Dynamics of Fluids in Porous Media,* Elsevier, New York, 527, 1972.

**Bedford, A., Costley, R. D., and Stern,** On the drag and virtual mass coefficients in Biot's equation, *J. Acoust. Soc. Am.,* 76, 1804, 1984.

**Berryman, J. G.,** Long-wavelength propagation in composite elastic media. I. Spherical inclusions, *J. Acoust. Soc. Am.,* 68, 1809, 1980a.

**Berryman, J. G.,** Long wavelength propagation in composite elastic media. II. Ellipsoidal inclusions, *J. Acoust. Soc. Am.,* 68, 1820, 1980b.

**Berryman, J. G.,** Confirmation of Biot's theory, *Appl. Phys. Lett.,* 37, 382, 1980c.

**Beydoun, W. B., Cheng, C. H., and Toksöz, M. N.,** Detection of subsurface permeable zones and fractures by means of tube waves, Society of Professional Well Log Analysts European Logging Symposium, 10th, London, U.K., Transactions, 12, 1983.

**Beydoun, W. B., Cheng, C. H., and Toksöz, M. N.,** Detection of open fractures with vertical seismic profiling, *J. Geophys. Res.,* 90, 4557, 1985.

**Bhasavanija, K.,** A Finite Difference Model of an Acoustic Logging Tool: The Borehole in a Horizontally Layered Geologic Medium, Ph.D. thesis, (T-2763), Colorado School of Mines, Golden, 1983.

**Billings, M. P., Fowler-Billings, K., Chapman, C. A., Chapman, R. W., and Goldthwait, R. P.,** The geology of the Mt. Washington Quadrangle, New Hampshire, Department of Resources and Economic Development, Concord, NH, 44, 1979.

**Biot, M. A.,** Propagation of elastic waves in a cylindrical bore containing a fluid, *J. Appl. Phys.,* 23, 997, 1952.

**Biot, M. A.,** Theory of elasticity and consolidaton for a porous ansotropic solid, *J. Appl. Phys.,* 26, 182, 1955.

**Biot, M. A.,** Theory of propagation of elastic waves in a fluid saturated porous rock. I. Low frequency range, *J. Acoust. Soc. Am.,* 28, 168, 1956a.

**Biot, M. A.,** Theory of propagation of elastic waves in a fluid saturated porous rock. II. Higher frequency range, *J. Acoust. Soc. Am.,* 28, 179, 1956b.

**Biot, M. A.,** General solutions of the equations of elasticity and consolidation for a porous material, *J. Appl. Mech.,* 78, 91, 1956c.

**Biot, M. A.,** Mechanics of deformation and acoustic propagation in porous media, *J. Appl. Phys.,* 33, 1482, 1962a.

**Biot, M. A.,** Generalized theory of acoustic propagation in porous dissipative media, *J. Acoust. Soc. Am.,* 34, 1254, 1962b.

**Biot, M. A.,** Nonlinear and semilinear rheology of porous solids, *J. Geophys. Res.,* 78, 4924, 1973.

**Biot, M. A. and Willis, D. G.,** The elastic coefficients of the theory of consolidation, *J. Appl. Mech.,* 24, 594, 1957.

**Birch, F.,** The velocity of compressional waves in rocks to 10 kb, Part 1, *J. Geophys. Res.,* 65, 1083, 1960.

**Birch, F.,** The velocity of compressional waves in rocks to 10 kb, Part 2, *J. Geophys. Res.,* 66, 2199, 1961.

**Birch, F.,** Compressibility elastic constants in *Handbook of Physical Constants, Geol. Soc. Am. Mem.,* 97, 97, 1966.

**Block, L. V., Cheng, C. H., and Duckworth, G. L.,** Velocity analysis of multi-receiver full waveform acoustic logging data in open and cased holes, *Log Anal.,* 32, 188, 1991.

**Boore, D. M.,** Finite difference methods for seismic waves, in *Methods in Computational Physics,* Vol. 11, Bolt, B. A., Ed., Academic Press, New York, 1, 1972.

**Bostrom, A. and Burden, A.,** Propagation of elastic surface waves along a cylindrical cavity and their excitation by a point force, *J. Acoust. Soc. Am.,* 72, 998, 1982.

**Bowles, J. E.,** *Engineering Properties of Soils and their Measurement,* McGraw-Hill, New York, 1978.

**Brace, W. F.,** Some new measurements of linear compressibility of rocks, *J. Geophys. Res.,* 70, 391, 1965.

**Brace, W. F.,** Pore pressure in geophysics, *Am. Geophys. Union Mono.,* 16, 265, 1972.

**Brace, W. F.,** Permeability from resistivity and pore shape, *J. Geophys. Res.,* 82, 3343, 1977.

**Brace, W. F. and Martin, R. J.,** A test of the laws of effective stress for crystalline rocks of low porosity, *Int. J. Rock Mech. Min. Sci.,* 5, 415, 1968.

**Brace, W. F., Walsh, J. B., and Frangos, W. T.,** Permeability of granite under high pressure, *J. Geophys. Res.,* 73, 2225, 1968.

**Brekhovskikh, L. M.,** *Waves in Layered Media,* Academic Press, New York, 361, 1960.

**Brown, R. J. S. and Korringa, J.,** On the dependence of the elastic properties of a porous rock on the compressibility of the pore fluid, *Geophysics,* 40, 608, 1975.

**Brune, J., Nafe, J., and Oliver, J.,** A simplified method for analysis and synthesis of dispersed wave trains, *J. Geophys. Res.,* 65, 287, 1960.

**Burns, D. R.,** Viscous fluid effects on guided wave propagation in a borehole, *J. Acoust. Soc. Am.,* 83, 463, 1988.

**Burns, D. R.,** Acoustic waveform logs and the *in situ* measurement of permeability, in *Geophysical Applications for Geotechnical Investigations,* Paillet, F. L. and Saunders, W. R., Eds., ASTM STP 1101, 65, 1989.

**Burns, D. R. and Cheng, C. H.,** Determination of in-situ permeability from tube wave velocity and attenuation, Society of Professional Well Log Analysts Annual Logging Symposium, 27th, Houston, Transactions, Paper KK, 1986.

**Burns, D. R. and Cheng, C. H.,** Inversion of borehole guided wave amplitudes for formation shear wave attenuation values, *J. Geophys. Res.,* 92, 12713, 1987.

**Burns, D. R., Cheng, C. H., Schmitt, D. P., and Toksöz, M. N.,** Permeability estimation from full waveform acoustic logging data, *Log Anal.,* 29, 112, 1988.

**Burridge, R. and Keller, J. B.,** Poroelasticity equations derived from microstructrure, *J. Acoust. Soc. Am.,* 70, 1140, 1981.

**Carroll, M. M.,** An effective stress law for anisotropic elastic deformation, *J. Geophys. Res.,* 84, 7510, 1979.

**Castagna, J. P., Batzle, M. L., and Eastwood, R. L.,** Relationships between compressional-wave and shear-wave velocities in clastic silicate rocks, *Geophysics,* 50, 571, 1985.

**Chan, A. K. and Tsang, L.,** Propagation of acoustic waves in a fluid filled borehole surrounded by a concentrically layered transversely isotropic formation, *J. Acoust. Soc. Am.,* 74, 1605, 1983.

**Chandler, R.,** Transient streaming potential measurements of fluid saturated porous structures: an experimental verification of Biot's slow wave in the quasistatic limit, *J. Acoust. Soc. Am.,* 70, 116, 1981.

**Chandler, R. and Johnson, D. L.,** The equivalence of quasi-static flow in fluid saturated porous media and Biot's slow wave in the limit of zero frequency, *J. Appl. Phys.,* 52, 3391, 1981.

**Chang, S. K. and Everhart, A. H.,** A study of sonic logging in a cased borehole, *J. Pet. Technol.,* 35, 1745, 1983.

**Chang, S. K., Liu, H. L., and Johnson, D. L.,** Low-frequency tube waves in permeable rocks, *Geophysics,* 53, 519, 1988.

**Charlaix, E., Kushnick, A. P., and Stokes, J. P.,** Experimental study of dynamic permeability in porous media, *Phys. Rev. Lett.,* 61, 14, 1595, 1988.

**Chen, S. T.,** The full acoustic wave train in a laboratory model of a borehole, *Geophysics,* 47, 1512, 1982.

**Chen, S. T.,** Shear-wave logging with dipole sources, *Geophysics,* 53, 659, 1988.

**Chen, S. T.,** Shear wave logging with multi-pole sources, *Geophysics,* 54, 590, 1989.

**Chen, S. T. and Willen, D. E.,** Shear wave logging in slow formation, Society of Professional Well Log Analysts Annual Logging Symposium, 25th, New Orleans, Transactions, Paper DD, 1984.

**Cheng, C. H.,** Seismic Velocities in Porous Rocks, Direct and Inverse Problems, Ph.D. thesis, Massachusetts Institute Technology, Cambridge, 1978.

**Cheng, C. H.,** Full waveform inversion of P waves for Vs and Qp, *J. Geophys. Res.,* 94, 19619, 1989.

**Cheng, C. H. and Johnston, D. H.,** Dynamic and static moduli, *Geophys. Res. Lett.,* 8, 39, 1981.

**Cheng, C. H. and Toksöz, M. N.,** Inversion of seismic velocities for the pore aspect ratio spectrum of a rock, *J. Geophys. Res.,* 84, 7533, 1979.

**Cheng, C. H. and Toksöz, M. N.,** Modelling of full wave acoustic logs, Society of Professional Well Log Analysts Annual Logging Symposium, 21st, Lafayette, LA, Transactions, Paper J, 1980.

**Cheng, C. H. and Toksöz, M. N.,** Elastic wave propagation in a fluid-filled borehole and synthetic acoustic logs, *Geophysics,* 46, 1042, 1981.

**Cheng, C. H. and Toksöz, M. N.,** Generation, propagation and analysis of tube wave in a borehole, Society of Professional Well Log Analysts Annual Logging Symposium, 23rd, Corpus Christi, TX, Transactions, Paper P, 1982.

**Cheng, C. H. and Toksöz, M. N.,** Determination of shear wave velocities in "slow" formation, Society of Professional Well Log Analysts Annual Logging Symposium, 24th, Calgary, Alberta, Canada, Transactions, Paper V, 1983.

**Cheng, C. H., Toksöz, M. N., and Willis, M. E.,** Velocity and attenuation from full waveform acoustic logs, Society of Professional Well Log Analysts Annual Logging Symposium, 22nd, Mexico City, Transactions, Paper O, 1981.

**Cheng, C. H., Toksöz, M. N., and Willis, M. E.,** Determination of *in situ* attenuation from full waveform acoustic logs, *J. Geophys. Res.,* 87, 5477, 1982.

**Cheng, C. H., Wilkens, R. H., and Meredith, J. A.,** Modelling of full waveform acoustic logs in soft marine sediments, Society Professional Well Log Analysts Annual Logging Symposium, 27th, Houston, TX, Transactions, Paper KK, 1986.

**Cheng, C. H., Zhang, J., and Burns, D. R.,** Effects of *in situ* permeability on the propagation of Stoneley (tube) waves in a borehole, *Geophysics,* 52, 1279, 1987.

**Christensen, N. I.,** Seismic velocities, in *Handbook of Physical Properties of Rocks,* Vol. 2, Carmichael, R. S., Ed., CRC Press, Boca Raton, FL, 1, 1982.

**Christiansen, D. M.,** A theoretical analysis of wave propagation in fluid-filled drill holes for the interpretation of three dimensional velocity logs, Society of Professional Well Log Analysts Annual Logging Symposium, 5th, Midland, TX, Transactions, Paper K, 1964.

**Clark, V. A., Spencer, T. W., Tittmann, B. R., Ahberg, L. A., and Coombe, L. T.,** Effect of volatiles on attenuation and velocity in sedimentary rocks, *J. Geophys. Res.,* 85, 5190, 1980.

**Crampin, S.,** An introduction to wave propagation in anisotropic media, *Geophys. J. R. Astron. Soc.,* 76, 29, 1984.

**Davison, C. C.,** Monitoring hydrogeological conditions in fractured rock at the site of Canada's Underground Research Laboratory, *Groundwater Monitoring Review,* Vol. 3, 95, 1984.

**DeWiest, R. J. M.,** *Flow Through Porous Media,* Academic Press, New York, 1969.

**Domenico, S. N.,** Elastic properties of unconsolidated porous sand reservoirs, *Geophysics,* 42, 1339, 1977.

**Duckworth, G. L.,** Processing and inversion of Arctic Ocean refraction data, Sc.D. thesis, Joint Program in Ocean Engineering, Massachusetts Institute Technology, Cambridge, and Woods Hole Oceanographic Institution, Woods Hole, MA, 1983.

**Dutta, N. C.,** Theoretical analysis of observed second bulk compressional wave in a fluid-saturated porous solid at ultrasonic frequencies, *Appl. Phys. Lett.,* 37, 898, 1980.

**Dutta, N. C. and Ode, H.,** Attenuation and dispersion of compressional waves in fluid filled porous rocks with partial gas saturation (White model). I. Biot's theory, *Geophysics,* 44, 1777, 1979a.

**Dutta, N. C. and Ode, H.,** Attenuation and dispersion of compressional waves in fluid filled porous rocks with partial gas saturation (White model). II. Results, *Geophysics,* 44, 1789, 1979b.

**Dutta, N. C. and Ode, H.,** Seismic reflections from a gas water contact, *Geophysics,* 48, 148, 1983.

**Ewing, W. N., Jardetsky, W. S., and Press, F.,** *Elastic Waves in Layered Media,* McGraw-Hill, New York, 357, 1957.

**Fatt, I.,** The compressibility of sandstones at low to moderate pressures, *Bull. Am. Assoc. Pet. Geol.,* 42, 1924, 1958.

**Fatt, I.,** The Biot-Willis elastic coefficients for a sandstone, *J. Appl. Mech.,* 26, 296, 1959.

**Feng, S., and Johnson, D. L.,** High frequency acoustic properties of a fluid/porous interface. I. New surface mode, *J. Acoust. Soc. Am.,* 74, 906, 1983a.

**Feng, S., and Johnson, D. L.,** High frequency acoustic properties of a fluid/porous interface. II. The 2D reflection Green's function, *J. Acoust. Soc. Am.,* 74, 915, 1983b.

**Feves, M. and Simmons, G.,** Effects of stress on cracks in westerly granite, *BSSA,* 66, 1755, 1976.

**Futterman, W. I.,** Dispersive body waves, *J. Geophys. Res.,* 67, 5257, 1962.

**Gardner, G. H. F., Wyllie, M. R. J., and Droschak, D. M.,** Effects of pressure and fluid saturation on the attenuation of elastic waves, *J. Pet. Technol.,* 16, 189, 1964.

**Garg, S. K. and Nur, A.,** Effective stress laws for fluid-saturated porous rocks, *J. Geophys. Res.,* 78, 5911, 1973.

**Gassmann, F.,** Elastic waves through a packing of spheres, *Geophysics,* 16, 673, 18, 269, 1951.

**Geertsma, J.,** The effect of fluid pressure decline on volumetric changes of porous rocks, *A.I.M.E. Trans.,* 210, 331, 1957.

**Geertsma, J. and Smit, D. C.,** Some aspects of elastic wave propagation in fluid-saturated porous solids, *Geophysics,* 26, 169, 1961.

**Geyer, R. L. and Myung, J. I.,** The 3-D velocity log — a tool for *in situ* determination of the elastic moduli of rocks in Chapter 4, Dynamic Rock Mechanics, U.S. Symposium on Rock Mechanics, 12th, Proceedings, 71, 1970.

**Goetz, J. F., Dupal, L., and Bowles, J.,** An investigation into the discrepancies between sonic log and seismic check shot velocities, *J. Aust. Pet. Explor. Assoc.,* 19, 131, 1979.

**Goldberg, D. and Gant, W. T.,** Shear-wave processing of sonic log waveforms in a limestone reservoir, *Geophysics,* 53, 668, 1988.

**Goldberg, D., Moos, D., and Anderson, R.,** Attenuation changes due to diagenesis in marine sediments: Society of Professional Well Log Analysts Annual Logging Symposium, 26th, Dallas, Transactions, Paper KK, 1985.

**Goldberg, D. and Zinszner, B.,** P-wave attenuation measurements from laboratory resonance and sonic waveform data, *Geophysics,* 54, 76, 1989.

**Grant, F. S. and West, G. F.**, *Interpretation Theory in Applied Geophysics*, McGraw Hill, New York, 584, 1960.

**Guyod, H. and Shane, L. E.**, *Geophysical Well Logging, Vol. 1, Introduction to Logging and Acoustic Logging*, Houston, Guyod, 256, 1969.

**Haddon, A. W.**, Numerical evaluation of Green's functions for axisymmetric boreholes using leaking modes, *Geophysics*, 52, 1099, 1987.

**Haddon, A. W.**, Exact Green's functions using leaking mode for axisymmetric boreholes in solid elastic media, *Geophysics*, 54, 609, 1989.

**Hadley, K.**, Comparison of calculated and observed crack densities and seismic velocities in westerly granite, *J. Geophys. Res.*, 81, 3484, 1976.

**Han, D. H., Nur, A., and Morgan, D.**, Effects of porosity and clay content on wave velocities in sandstones, *Geophysics*, 51, 2093, 1986.

**Handin, J., Hager, R. V., Jr., Friedman, M., and Feather, J. N.**, Experimental deformation of sedimentary rocks under confining pressure: pore pressure test, *Bull. Am. Assoc. Pet. Geol.*, 47, 717, 1963.

**Hardin, E. L., Cheng, C. H., Paillet, F. L., and Mendelson, J. D.**, Fracture characterization by means of attenuation and generation of tube waves in fractured crystalline rock at Mirror Lake, New Hampshire, *J. Geophys. Res.*, 92, 7989, 1987.

**Hardin, E. L. and Toksöz, M. N.**, Detection and characterization of fractures from generation of tube waves. Society of Professional Well Log Analysts Annual Logging Symposium, 26th, Dallas, Transactions, Paper II, 1985.

**Haskell, N. A.**, The dispersion of surface waves in multilayered media, *Bull. Seismol. Soc. Am.*, 43, 17, 1953.

**Hearst, J. R. and Nelson, P. H.**, *Well Logging for Physical Properties*, McGraw-Hill, New York, 570, 1985.

**Hornby, B. E., Johnston, D. L., Winkler, K. H., and Plumb, R. A.**, Fracture evaluation from the borehole Stoneley wave, *Geophysics*, 54, 1274, 1989.

**Hovem, J. M.**, Viscous attenuation of sound in suspensions and high porosity sediments, *J. Acoust. Soc. Am.*, 67, 1559, 1980.

**Hsieh, P. A., Neuman, S. P., and Simpson, E. S.**, Pressure testing of fractured rocks — a methodology employing three-dimensional hole test, U.S. Nuclear Reg. Comm., Washington, D.C., NUREG/CR-3213, 176, 1983.

**Hsu, K. and Baggeroer, A. B.**, Application of the maximum-likelihood method (MLM) for sonic velocity logging, *Geophysics*, 51, 780, 1986.

**Hsui, A. T., Zhang, J., Cheng, C. H., and Toksöz, M. N.**, Tube wave attenuation and *in situ* permeability, Society of Professional Well Log Analyst Annual Logging Symposium, 26th, Dallas, Transactions, Paper CC, 1985.

**Hsui, A. T. and Toksöz, M. N.**, Application of an acoustic model determine *in situ* permeability of a borehole, *J. Acoust. Soc. Am.*, 79, 2055, 1986.

**Huang, C. F. and Hunter, J. A.**, The correlation of "tube wave" events with open fractures in fluid filled boreholes, Current Research, Part A. Geological Survey of Canada, Paper 81-1A, 361, 1981a.

**Huang, C. F. and Hunter, J. A.**, A seismic tube wave method for the *in situ* estimation of rock fracture permeability in boreholes, Soc. Expl. Geophys. Annual Int. Meet., 51st, Tech. Prog. Abst., 414, 1981b.

**Hubbert, M. K. and Rubey, W. W**, 1959, Role of fluid pressure in mechanics of overthrust faulting, 1, 2, *Bull. Geol. Soc. Am.*, 70, 115, 1959.

**Hughes, D. S. and Cooke, C. E.**, The effect of pressure on the reduction of pore volume of consolidated sandstones, *Geophysics*, 18, 298, 1953.

**Hughes, D. S. and Cross, J. H.**, Elastic wave velocities in rocks at high pressures and temperatures, *Geophysics*, 16, 577, 1951.

**Jacobi, W. J.**, Propagation of sound waves along liquid cylinders, *J. Acoust. Soc. Am.*, 21, 120, 1949.

**Johnson, D. L., Plona, T. J., Scala, C., Pasierb, F., and Kojima, F.**, Tortuosity and acoustic slow waves, *Phys. Rev. Lett.*, 49, 1840, 1982.

**Johnston, D. H.,** The Attenuation of Seismic Waves in Dry and Saturated Rocks, Ph.D. thesis, Massachusetts Institute of Technology, Cambridge, 78, 1978.

**Johnston, D. H.,** Physical properties of shale at temperature and pressure, *Geophysics,* 52, 1391, 1987.

**Johnston, D. H. and Toksöz, M. N.,** Ultrasonic P and S wave attenuation in dry and saturated rocks under pressure, *J. Geophys. Res.,* 85, 925, 1980.

**Johnston, D. H. and Toksöz, M. N.,** Definitions and terminology in Seismic Wave Attenuation, *Geophysics Reprint Series, No. 2,* Toksöz, M. N. and Johnston, D. H., Eds., Society of Explor. Geophysicists, Tulsa, OK, 1981.

**Johnston, D. H., Toksöz, M. N., and Timur, A.,** The attenuation of seismic waves in dry and saturated rocks. II. Mechanisms, *Geophysics,* 44, 691, 1979.

**Jones, T. D.,** Pore fluids and frequency-dependent wave propagation in rocks, *Geophysics,* 51, 1939, 1986.

**Kaneko, F., Kanemori, T., and Tonouchi, K.,** Low-frequency shear wave logging in unconsolidated formations for geotechnical applications, in *Geophysical Applications for Geotechnical Investigations,* Paillet, F. L. and Saunders, W. R., Eds., ASTM STP 1101, 79, 1989.

**Katsube, T. J. and Hume, J. P.,** Permeability determination in crystalline rocks by standard geophysical logs, *Geophysics,* 52(3), 342, 1987.

**Keys, W. S.,** Borehole geophysics in igneous and metamorphic rocks: Society of Professional Well Log Analyst Annual Logging Symposium, 20th, Tulsa, OK, Transactions, Paper OO, 1979.

**Kelly, K. R., Ward, R. W., Treitel, S., and Alford, R. M.,** Synthetic seismograms: a finite difference approach, *Geophysics,* 41, 2, 1976.

**Kimball, C. V. and Marzetta, T. L.,** Semblance processing of borehole acoustic array data, *Geophysics,* 49, 274, 1984.

**Khilar, K. C. and Fogler,** Water sensitivity of sandstones, *J. Soc. Pet. Eng.,* AIME, 55, 1983.

**King, M. S.,** Wave velocities in rocks as a function of changes in overburden pressure and pore fluid saturant, Geophysics, 31, 50, 1966.

**King, M. S.,** Ultrasonic compressional and shear-wave velocities of confined rock samples, in *Canadian Rock Mechanics Symposium,* 5th, Hughes, J., Ed., Toronto, Canada, Proceedings, 127, 1968.

**King, M. S.,** Static and dynamic moduli of rocks under pressure, in *Rock Mechanics — Theory and Practice: Symposium on Rock Mechanics,* 11th, Somerton, W. H., Ed., University of California, Berkeley, Proceedings, 329, 1969.

**Kitsunezaki, C.,** A new method for shear wave logging, *Geophysics,* 45, 1489, 1980.

**Kjartansson, E.,** Constant Q wave propagation and attenuation, *J. Geophys. Res.,* 84, 4734, 1979.

**Knutson, C. F. and Bohor, B. F.,** Reservoir rock behavior under moderate confining pressure, in 5th Symposium on Rock Mechanics, University of Minnesota, Proceedings, 627, 1963.

**Koerperich, E. A.,** Investigation of acoustic boundary waves and interference patterns as techniques for detecting fractures, *J. Petro. Technol.,* 30, 1199, 1978.

**Koerperich, E. A.,** Shear wave velocities determined from long and short spaced borehole acoustic devices, *Soc. Pet. Engr.,* Paper SPE 8237, 12, 1979.

**Koerperich, E. A.,** Shear wave velocities determined from long and short spaced borehole acoustic devices, *Soc. Pet. Engr. J.,* 20, 317, 1980.

**Korringa, J., Brown, R. J. S., Thompson, D. D., and Runge, R. J.,** Self-consistent imbedding and the ellipsoidal model for porous rocks, *J. Geophys. Res.,* 84, 5591, 1979.

**Kurkjian, A. L.,** Farfield decomposition of acoustic waveforms in a fluid-filled borehole, *J. Acoust. Soc. Am.,* 74, suppl., 88, 1983.

**Kurkjian, A. L.,** Numerical computation of individual far field arrivals excited by an acoustic source in a borehole, *Geophysics,* 50, 852, 1985.

**Kurkjian, A. L.,** Theoretical far-field radiation from a low-frequency horizontal point force in a vertical borehole, *Geophysics,* 51, 930, 1986.

**Kurkjian, A. L. and Chang, S. K.,** Acoustic multipoles in fluid-filled boreholes, *Geophysics,* 51, 148, 1986.

**Kuster, G. T. and Toksöz, M. N.,** Velocity and attenuation of seismic waves in two phase media. I. Theoretical formulation, *Geophysics,* 39, 587, 1974a.

**Kuster, G. T. and Toksöz, M. N.,** Velocity and attenuation of seismic waves in two phase media. II. Experimental results, *Geophysics,* 39, 607, 1974b.

**Lebreton, F., Sarda, J. P., Trocqueme, F., and Molier, P.,** Logging tests in porous media to evaluate the influence of their permeability on accoustic waveforms, Society of Professional Well Log Analysts Annual Logging Symposium, 19th, El Paso, TX, Transactions, Paper Q, 1978.

**Lindseth, R. O.,** Synthetic sonic logs — A process for stratigraphic interpretation, *Geophysics,* 44, 3, 1979.

**Liu, O. Y.,** Stoneley wave-derived Dt shear log, Society of Professional Well Log Analysts Annual Logging Symposium, 25th, New Orleans, Transactions, Paper ZZ, 1984.

**Lo, Tien-when, Coyner, K. B., and Toksöz, M. N.,** Experimental determination of elastic ansiotrophy of Berea sandstone, Chicopee shale, and Chelmsford granite, *Geophysics,* 51, 164, 1986.

**Love, A. E. H.,** *A Treatise on the Mathematical Theory of Elasticity,* Dover, New York, 1944.

**Mair, J. A. and Green, A. G.,** High resolution seismic reflection profiles reveal fractures within a homogeneous granite batholith, *Nature,* 294, 439, 1981.

**Mann, R. L. and Fatt, I.,** Effect of pore fluids on the elastic properties of sandstones, *Geophysics,* 25, 433, 1960.

**Mathieu, F.,** Application of Full Waveform Logging Data to the Estimation of Reservoir Permeability, Master's thesis, Massachusetts Institute of Technology, Cambridge, 69, 1984.

**Mathieu, F. and Toksöz, M. N.,** Application of full waveform acoustic logging data to the estimation of reservoir permeability, Soc. Expl. Geophys. Annu. Int. Meet., 54th, Proceedings, Atlanta, 9, 1984.

**Mavko, G. M and Nur, A.,** The effect of non elliptical cracks on the compressibility of rocks, *J. Geophys. Res.,* 83, 4459, 1978.

**Mavko, G. M and Nur, A.,** Wave attenuation in partially saturated rocks, *Geophysics,* 44, 161, 1979.

**McGarr, A. and Gay, N. C.,** State of stress in the earth's crust, *Annu. Rev. Earth Planet. Sci.,* 6, 405, 1978.

**Moos, D. and Zoback, M. D.,** *In situ* studies of velocity in fractured crystalline rocks, *J. Geophys. Res.,* 88, 2345, 1983.

**Morse, P. M.,** The transmission of sound inside pipes, *J. Accoust. Soc. Am.,* 1, 205, 1939.

**Morse, P. M. and Feshback, H.,** *Methods of Theoretical Physics,* McGraw-Hill, New York, 1978, 1953.

**Morris, R. L., Grine, D. L., and Arkfield, T. E.,** Using compression and shear acoustic amplitudes for the location of fractures, *J. Pet. Technol.,* 13, 623, 1965.

**Murphy, W. F.,** Acoustic measures of partial gas saturation in tight sandstones, *J. Geophys. Res.,* 89, 11549, 1984.

**Norris, A. N.,** Stoneley-wave attenuation and dispersion in permeable formations, *Geophysics,* 54, 330, 1989.

**Nur, A. and Byerlee, J. D.,** An exact effective stress law for elastic deformation of rocks with fluids, *J. Geophys. Res.,* 76, 6414, 1971.

**Nur, A. and Byerlee, J. D.,** Seismic velocities in dry and saturated cracked solids, *J. Geophys. Res.,* 79, 5412, 1974.

**Nur, A. and Byerlee, J. D.,** Viscoelastic properties of fluid-saturated cracked solids, *J. Geophys. Res.,* 82, 5719, 1977.

**Nur, A., and Simmons, G.,** The effect of saturation on velocity in low porosity rocks, *Earth Planet Sci. Lett.,* 7, 183, 1969.

**Ogushwitz, P. R.,** Applicability of the Biot theory. I. Low-porosity materials, *J. Acoust. Soc. Am.,* 77, 429, 1985.

**Paillet, F. L.,** Acoustic propagation in the vicinity of fractures which intersect a fluid-filled borehole, Society of Professional Well Log Analysts Annual Logging Symposium, 21st, Lafayette, LA, Transactions, Paper DD, 1980.

**Paillet, F. L.,** Predicting the frequency content of acoustic waves in boreholes, Society of Professional Well Log Analysts Annual Logging Symposium, 22nd, Mexico City, Transactions, Paper SS, 1981a.

**Paillet, F. L.,** A comparison of fracture characterization techniques applied to near vertical fractures in a limestone reservoir, Society of Professional Well Log Analysts Annual Logging Symposium, 22nd, Mexico City, Transactions, Paper XX, 1981b.

**Paillet, F. L.,** Acoustic characterization of fracture permeability at Chalk River, Ontario, Canada, *Can. Geotech. J.,* 20, 468, 1983a.

**Paillet, F. L.,** Frequency and scale effects in the optimization of acoustic waveform logs, Society of Professional Well Log Analysts Annual Logging Symposium, Calgary, Alberta, Canada, 24th, Transactions, Paper U, 1983b.

**Paillet, F. L.,** Field test of a low-frequency sparker source for acoustic waveform logging, Society of Professional Well Log Analysts Annual Logging Symposium, 25th, New Orleans, Transactions, Paper GG, 1984.

**Paillet, F. L.,** Problems in fractured-reservoir evaluation and possible routes to their solution, *Log Anal.,* 26, 26, 1985.

**Paillet, F. L.,** Fracture characterization and fracture-permeability estimation at the Underground Research Laboratory in Southeastern Manitoba, Canada, U.S. Geological Survey Water Resources Investigations Report 88-4009, 42, 1988.

**Paillet, F. L.,** Qualitative and quantitative interpretation of fracture permeability by means of acoustic full waveform logs, *Log Anal.,* 32, 256, 1991.

**Paillet, F. L., Barton, C. A., Luthi, S.M., Rambow, F.H.K., and Zemanek, J.,** Borehole imaging, Houston, Society of Professional Well Log Analysts, 472, 1990.

**Paillet, F. L. and Cheng, C. H.,** A numerical investigation of head waves and leaky modes in fluid-filled boreholes, *Geophysics,* 51, 1438, 1986.

**Paillet, F. L., Cheng, C. H., and Meredith, J. A.,** New applications in the inversion of acoustic full waveform logs — relating mode excitation to lithology, *Log Anal.,* 28, 307, 1987.

**Paillet, F. L., Cheng, C. H., and Tang, X. M.,** Theroetical models relating acoustic tube-wave attenuation to fracture permeability — reconciling model results with field data, Society of Professional Well Log Analysts Annual Logging Symposium, 30th, Denver, Transactions, Paper FF, 1989.

**Paillet, F. L. and Hess, A. E.,** Geophysical well log analysis of fractured crystalline rocks at East Bull Lake, Ontario, Canada, U.S. Geological Survey Water-Resources Investigations Report 86-4052, 1986.

**Paillet, F. L. and Hess, A. E.,** Geophysical well log analysis of fractured granitic rocks at Atikokan, Ontario, Canada, U.S. Geological Survey Water Resources Investigations Report 87-4154, 36, 1987.

**Paillet, F. L. and Hess, A. E.,** Characterizing flow paths and permeability distribution in fractured-rock aquifers using a sensitive, thermal borehole flowmeter, Molz, F. J., Melville, J. G. and Guven, O., Eds., New Field Techniques for Quantifying the Physical and Chemical Properties of Heterogeneous Aquifers Conference, Dallas, Proceedings, 445, 1989.

**Paillet, F. L., Hess, A. E., Cheng, C. H., and Hardin, E. L.,** Characterization of fracture permeability with high-resolution vertical flow measurements during borehole pumping, *Ground Water,* 25, 28, 1987.

**Paillet, F. L. and White, J. E.,** Acoustic modes of propagation in the borehole and their relationship to rock properties, *Geophysics,* 47, 1215, 1982.

**Pandit, B. I. and King, M. S.,** The variation of elastic wave velocities and quality factor of a sandstone with moisture content, *Can. J. Earth Sci.,* 16, 2187, 1979.

**Pascal, H.,** Pressure wave propagation in a fluid flowing through a porous medium and problems related to interpretation of Stoneley's wave attenuation in acoustic well logging, *Int. J. Eng. Sci.,* 24, 1553, 1986.

**Peterson, E. W.,** Acoustic wave propagation along a fluid-filled cylinder, *J. Appl. Phys.,* 45, 3340, 1974.

**Phinney, R. A.,** Leaking modes in the crustal waveguide, Part 1. The ocean PL wave, *J. Geophys. Res.,* 65, 1445, 1961.

**Pickens, J. F., Grisak, G. E., Avis, J. D., Belanger, D. W., and Thury, M.,** 1987, Analysis and interpretation of borehole hydraulic tests in deep boreholes — Principles, model development, and applications, *Water Resour. Res.,* 23, 1341, 1987.

**Pickett, G. R.,** The use of acoustic logs in the evaluation of sandstone reservoirs, *Geophysics,* 25, 250, 1960.

**Pickett, G. R.,** Acoustic character logs and their application information evaluation, *J. Pet. Technol.,* 15, 659, 1963.

**Plona, T. J.,** Observation of second bulk compressional wave in a porous medium at ultrasonic frequencies, *Appl. Phys. Lett.,* 36, 259, 1980.

**Poeter, E. P.,** Characterizing fractures at potential nuclear waste repository sites with acoustic waveform logs, *Log Anal.,* 28, 453, 1987.

**Postma, G. W.,** Wave propagation in a stratified medium, *Geophysics,* 20, 780, 1955.

**Rayleigh, L.,** On waves propagated along the plane surface of an elastic solid, *Proc. London Math. Soc.,* 17, 4, 1885.

**Roever, W. L., Vining, T. F., and Strick, E.,** Propagation of elastic wave motion from an impulsive source along a fluid/solid interface, *Philos. Trans. R. Soc. London,* A251, 455, 1959.

**Roever, W. L., Rosenbaum, J., and Vining, T. F.,** Acoustic waves from an impulsive source in a fluid-filled borehole, *J. Acoust. Soc. Am.,* 55, 1144, 1974.

**Rosenbaum, J. H.,** The long time response of a layered elastic medium to explosive sound, *J. Geophys. Res.,* 65, 1577, 1960.

**Rosenbaum, J. H.,** Synthetic microseismograms: logging in porous formations, *Geophysics,* 39, 14, 1974.

**Schmitt, D. P.,** Shear wave logging in elastic formations, *J. Acoust. Soc. Am.,* 84, 2215, 1988.

**Schmitt, D. P. and Bouchon, M.,** Full wave acoustic logging, Synthetic microseismograms and frequency wavenumber analysis, *Geophysics,* 50, 1756, 1985.

**Schmitt, D. P., Bouchon, M., and Bonnet, G.,** Full-wave synthetic acoustic logs in radially semiinfinite saturated porous media, *Geophysics,* 53, 807, 1988.

**Schmitt, D. P., Zhu, Y., and Cheng, C. H.,** Shear wave logging in semi-infinite saturated porous formations, *J. Acoust. Soc. Am.,* 84, 2230, 1988.

**Schoenberg, M., Marzetta, T., Aron, J., and Porter, R.,** Space time dependence of acoustic waves in a borehole, *J. Acoust. Soc. Am.,* 70, 1496, 1981.

**Schoenberg, M., Sen, P. N., and White, J. E.,** Attenuation of acoustic modes due to viscous drag at the borehole wall, *Geophysics,* 52, 1566, 1987.

**Seeburger, D. A. and Nur, A.,** A pore space model for rock permeability and bulk modulus, *J. Geophys. Res.,* 89, 527, 1984.

**Simmons, G.,** Velocity of shear waves in rocks to 10 kilobars, 1, *J. Geophys. Res.,* 69, 1123, 1964.

**Simmons, G., Siegfried, R. W., and Feves, M.,** Differential strain analysis: a new method for examining cracks in rocks, *J. Geophys. Res.,* 79, 4383, 1974.

**Snow, D. T.,** A parallel plate model of fractured permeable media, Ph.D. thesis, University of California, Berkeley, 78, 1965.

**Somers, E. V.,** Propagation of acoustic waves in a liquid filled cylindrical hole, surrounded by an elastic solid, *J. Appl. Phys.,* 24, 515, 1953.

**Spencer, J. W.,** Bulk and shear attenuation in Berea sandstone: the effects of pore fluids, *J. Geophys. Res.,* 84, 7521, 1979.

**Spencer, J. W.,** Stress relaxations at low frequencies in fluid-saturated rocks — attenuation and modulus dispersion, *J. Geophys. Res.,* 86, 1803, 1981.

**Spencer, J. W. and Nur, A. M.,** The effects of pressure, temperature, and pore water on velocities in westerly granite, *J. Geophys. Res.,* 81, 899, 1976.

**Stephen, R. A.,** A comparison of finite difference and reflectivity seismograms for marine models, *Geophys. J. R. Astron. Soc.,* 72, 39, 1983.

**Stephen, R. A., Pardo-Casas, F., and Cheng, C. H.,** Finite difference synthetic acoustic logs, *Geophysics,* 50, 1588, 1985.

Stevens, J. L. and Day, S. M., Shear velocity logging in slow formations using the Stoneley wave, *Geophysics*, 51, 137, 1986.

Stewart, R. R., Huddleston, P. D., and Kan, T. K., Seismic versus sonic velocities: a vertical seismic profiling study, *Geophysics*, 49, 1153, 1984.

Stierman, D. J. and Kovach, R. L., An *in situ* velocity study — the Stone Canyon Well, *J. Geophys. Res.*, 84, 672, 1979.

Stoll, R. D., Acoustic waves in saturated sediments, in *Physics of Sound in Marine Sediments*, Hampton, L., Ed., Plenum Press, New York, 19, 1974.

Stoll, R. D. and Bryan, G. M., Wave attenuation in saturated sediments, *J. Acoust. Soc. Am.*, 47, 1440, 1970.

Stoneley, R., The seismological implications of aeolotropy in continental structure, *Mon. Not. R. Astron. Soc., Geophys. Suppl.*, 5, 343, 1949.

Strick, E., An explanation of observed time discrepancies between continuous and conventional well surveys, *Geophysics*, 36, 285, 1971.

Summers, G. C. and Broding, R. A., Continuous velocity logging, *Geophysics*, 17, 598, 1952.

Tang, X. M. and Cheng, C. H., A dynamic model for fluid flow in open borehole fractures, *J. Geophys. Res.*, 94, 7567, 1989.

Tang, X.M., Cheug, C.H., and Paillet, F.L., Modeling borehole stoneley wave propagation across permeable *in situ* fracture, Society of Professional well Log Analysts Annual Logging Symposium, 32nd, Midland, Transactions, Papu 66, 1991.

Tang, X.M., Toksöz, M.N., and Cheug, C.H., Elastic wave radiation and diffraction of a piston source, J. Acoust. Sco. An., 87, 1894, 1990.

Tatham, R. H., Vp/Vs and lithology, *Geophysics*, 47, 326, 1982.

Thomas, D. H., Seismic applications of sonic logs, *Log Anal.*, 19, 23, 1978.

Thomsen, L., Weak elastic anisotropy, *Geophysics*, 51, 1954, .

Timoshenko, S. and Goodier, J. N., *Theory of Elasticity*, McGraw-Hill, New York, 506, 1951.

Timur, A., Temperature dependence on compressional and shear wave velocities in rocks, *Geophysics*, 42, 950, 1977.

Timur, A., Hempkins, W. B., and Weinbrandt, R. M., Scanning electron microscope study of pore systems in rocks, *J. Geophys. Res.*, 76, 4932, 1971.

Tittmann, B. R., Internal friction measurements and their implications in seismic Q structure models of the crust, in *Seismic Wave Attenuation*, Johnston, D.H. and Toksöz, M.N., Eds., Society of Explor. Geophysicists, Tulsa, OK, 81, 1981.

Tittmann, B. R., Clark, V. A., and Richardson, J. M., Possible mechanisms for seismic attenuation in rocks containing small amounts of volatiles, *J. Geophy. Res.*, 85, 5199, 1980.

Tittmann, B. R., Nadler, H., Clark, V. A., and Ahlberg, L. A., Frequency dependence of seismic dissipation in saturated rocks, *Geophys. Res. Lett.*, 8, 36, 1981.

Tittmann, J., *Geophysical Well Logging*, Academic Press, Orlando, 175, 1986.

Toksöz, M. N., Cheng, C. H., and Timur, A., Velocities of seismic waves in porous rocks, *Geophysics*, 41, 621, 1976.

Toksöz, M. N. and Cheng, C. H., Modeling of seismic velocities in porous rock and its application to seismic exploration, *Arab. J. Sci. Eng.*, Special Issue, 1978.

Toksöz, M. N., Johnston, D. H., and Timur, A., Attenuation of seismic waves in dry and saturated rocks. I. Laboratory measurements, *Geophysics*, 44, 681, 1979.

Toksöz, M. N. and Johnston, D. H., Eds., Seismic wave attenuation, S.E.G. Geophys. Reprint Ser., 2, Tulsa, OK, 459, 1981.

Toksöz, M. N., Wilkens, R. H., and Cheng, C. H., Shear wave velocity and attenuation in ocean bottom sediments from acoustic log waveforms, *Geophys. Res. Lett.*, 12, 37, 1985.

Tongtaow, C., Wave propagation along a cylindrical borehole in a transversely isotropic medium, Ph.D. thesis, Colorado School of Mines, Golden, 78, 1982.

Tosaya, C. and Nur, A., Effects of diagenesis and clays on compressional velocities in rocks, *Geophys. Res. Lett.*, 9, 5, 1982.

Tsang, L. and Rader, D., Numerical evaluation of transient acoustic waveform due to a point source in a fluid-filled borehole, *Geophysics*, 44, 1706, 1979.

Tubman, K. M., Cheng, C. H., and Toksöz, M. N., Synthetic full waveform acoustic logs in cased boreholes, *Geophysics,* 49, 1051, 1984.

Tubman, K. M., Cheng, C. H., Cole, S. P., and Toksöz, M. N., Synthetic full-waveform acoustic logs in cased boreholes, II — poorly bonded casing, *Geophysics,* 51, 902, 1986.

Vogel, C. B., A seismic velocity logging method, *Geophysics,* 17, 586, 1952.

Vogel, C. B. and Herolz, R. A., The CAD, a circumferential acoustical device for well logging, Society of Petroleum Engineers of AIME, 52nd, Paper 6819, 1977.

Walsh, J. B., The effect of cracks on the compressibility of rock, *J. Geophys. Res.,* 70, 381, 1965.

Walsh, J. B., Effect of pore pressure and confining pressure on fracture permeability, *Int. J. Rock Mech. Min. Sci.,* 18, 1981.

White, J. E., Elastic waves along a cylindrical bore, *Geophysics,* 27, 327, 1962.

White, J. E., *Seismic Waves: Transmission, Radiation and Attenuation,* McGraw-Hill, New York, 380, 1965.

White, J. E., The hula log-a proposed acoustic tool, Society of Professor Well Log Analysts Annual Logging Symposium, 8th, Denver, Transactions, Paper I, 1967.

White, J. E., Computed waveforms in transversely isotropic media, *Geophysics,* 47, 771, 1982.

White, J. E., *Underground Sound-Applications of Seismic Waves,* New York, Elsevier, 249, 1983.

White, J. E. and Angona, F. A., Elastic wave velocities in laminated media, *J. Acoust. Soc. Am.,* 27, 310, 1955.

White, J. E., Martineau-Nicoletis, L., and Monash, C., Measured anisotropy in Pierre Shale, *Geophys. Prospect.,* 31, 709, 1983.

White, J. E. and Sengbush, R. L., Velocity measurements in near-surface formations, *Geophysics,* 18, 54, 1953.

White, J. E. and Tongtaow, C., Cylindrical waves in transversely isotropic media, *J. Acoust. Soc. Am.,* 70, 1147, 1981.

White, J. E. and Zechman, R. E., Computed response of an acoustic logging tool, *Geophysics,* 33, 302, 1968.

Wilkens, R. H., Simmons, G., and Caruso, L., The ratio Vp/Vs as a discriminant of composition for siliceous limestones, *Geophysics,* 49, 1850, 1984.

Williams, D. M., Zemanek, J., Angona, F. A., Dennis, C. L., and Caldwell, R. L., The long space acoustic logging tool, Society of Professional Well Log Analyst Annual Logging Symposium, 25th, New Orleans, Transactions, Paper T, 1984.

Willis, M. E., Seismic Velocity and Attenuation from Full Waveform Acoustic Logs, Ph.D. thesis, Massachusetts Institute of Technology, Cambridge, 1983.

Willis, M. E. and Toksöz, M. N., Automatic P and S velocity determination from full waveform acoustic logs, *Geophysics,* 48, 1631, 1983.

Winbow, G. A., Compressional and shear arrivals in a multipole sonic log, *Geophysics,* 50, 1119, 1985.

Winbow, G. A., A theoretical study of acoustic S-wave and P-wave velocity logging with conventional and dipole sources in soft formations, *Geophysics,* 53, 1334, 1988.

Winkler, K. W., Estimates of velocity dispersion between seismic and ultrasonic frequencies, *Geophysics,* 51, 183, 1986.

Winkler, K. W., Liu, Hsui-Lin, and Johnson, D. L., Permeability and borehole Stoneley waves — comparison between experiment and theory, *Geophysics,* 54, 66, 1989.

Winkler, K. W. and Nur, A., Pore fluids and seismic attenuation in rocks, *Geophys. Res. Lett.,* 6, 1, 1979a.

Winkler, K. W. and Nur, A., Friction and seismic attenuation rocks, *Nature,* 277, 528, 1979b.

Winkler, K. W. and Nur, A., Seismic attenuation — effect of pore fluids and frictional sliding, *Geophysics,* 47, 1, 1982.

Winkler, K. W. and Plona, T. J., Technique for measuring ultrasonic velocity and attenuation spectra in rocks under pressure, *J. Geophys. Res.,* 87, 10776, 1982.

Witherspoon, P. A., Tsang, Y. W., Long, J. C. S., and Jahandar, N., New approaches to problems of fluid flow in fractured rock masses, Rock Mech. Symp., 22nd, Proceedings, Cambridge, MA, 3, 1981.

Wood, A. B., *A Textbook of Sound*, G. Bell and Sons, Ltd., London, 1941.

Wyllie, M. R. J., Gardner, G. H. F., and Gregory, A. R., Studies of elastic wave attenuation, *Geophysics*, 27, 569, 1962.

Wyllie, M. R. J., Gregory, A. R., and Gardner, L. W., Elastic wave velocities in heterogeneous and porous media, *Geophysics*, 21, 41, 1956.

Wyllie, M. R. J., Gregory, A. R., and Gardner, G. H. F., An experimental investigation of factors affecting elastic wave velocities in porous media, *Geophysics*, 23, 459, 1958.

Young, T. K., Application of generalized ray theory to the study of elastic wave propagation in the borehole environment, Ph.D. thesis, Colorado School of Mines, Golden, 118, 1979.

Zemanek, J., Jr. and Rudnick, I., Attenuation and despersion of elastic waves in a cylindrical bar, *J. Acoust. Soc. Am.*, 33, 1283, 1961.

Zemanek, J., Caldwell, R. L., Glenn, E. E., Halcomb, S. V., Norton, L. J., and Strauss, A. J. D., The borehole televiewer — a new logging concept for fracture location and other types of borehole inspection, *J. Pet. Technol.*, 21, 762, 1969.

Zemanek, J. Glenn, E. E., Norton, L. J., and Caldwell, R. L., Formation evaluation by inspection with the borehole televiewer, *Geophysics*, 35, 254, 1970.

Zemanek, J., Angona, F. A., Williams, D. M., and Caldwell, R. L., Continuous acoustic shear wave logging, Society of Professional Well Log Analyst Annual Logging Symposium, 25th, New Orleans, Transactions, Paper U, 1984.

Zhang, J. and Cheng, C. H., Numerical studies of body wave amplitudes in full waveform acoustic logs, Trans. 9th Eur. Formation Evaluation Symp., Paper 14, 1984.

Ziegler, T. W., Determination of rock mass permeability: U.S. Army Engineers Waterways Exp. Station, Vicksburg, Tech. Report S-76-2, 1976.

Zlatov, P., Poeter, E., and Higgins, J., Physical modeling of the full acoustic waveform in a fractured, fluid-filled borehole, *Geophysics*, 53, 1219, 1988.

# Index

## A

Abnormal geopressures, 26
Acoustic transit-time logs, 1, 2
Acoustic velocities
  abnormal geopressure indicators, 26
  soft materials and, 39
Airy phase, 76, 173
Amplitudes
  attenuation determination, 139
  head wave, 51
  P wavetrain vs. pseudo-Rayleigh
    mode, 125, 126
  spectral, cased boreholes, 183, 184
Amplitude threshold setting, 23
Angle of refraction, critical, 39, 40
Arrival picks, 23
Arrival time/moveout velocity plots, 168,
  171, 174, 176, 178, 180
Aspect ratio, 13
Attenuated modes, types of, 57, 58, see also
  PL modes; PT modes
Attenuation ($Q$), 42
  dilatational and shear waves, 198, 199
  fluid-filled boreholes, 81–89
  in hard formations, 112, 113
  intrinsic, 8–10, 84
  leaky P mode, 138
  laboratory measurements, 13–16
  lithology and, 117–121
  natural, 30, 81, 82
  in open boreholes, 138–145
    full waveform inversion of P wavetrain,
      143–145
    partition coefficients, 139–143
    waveform amplitudes, 139
  in soft formations, 115
  tube wave, 224–227
  viscous, 9, 10
Axisymmetric forcing, 27
Axisymmetric pressure field
  boundary condition matrix
    elements, 99–104
  fluid-filled boreholes, polarized shear
    waves, 89, 90
  point source production, 79
Axisymmetric propagation, 19
Azimuthal periodicity, 77, 78

## B

Beam spreading, 14
Biot model, 35, 191–197
Bonding, cased boreholes, see Cased
  boreholes
Borehole response function, fluid-filled
  boreholes, 83
Boreholes, cased, see Cased boreholes
Boreholes, open, see Open boreholes
Boundary condition matrices
  fluid-filled boreholes
    forcing with axisymmetric pressure
      sources, 99, 100, 103, 104
    forcing in transversely isotropic
      formations, 103, 104
    forcing with nonaxisymmetric pressure
      sources, 101–104
  fluid-filled boreholes, porous formations
    axisymmetric modes, 214, 215
    nonaxisymmetric modes, 216–218
Boundary conditions, 79
  cased boreholes, 151, 187, 188
  fluid-filled boreholes, 47, 48, 68, 70–75
    derivation of period equations for
      trapped modes, 71–75
    matrices, see Boundary condition
      matrices
  and wave equations, 41, 42
Branch cut integrals, 42, 43
  attenuation measurements, 144
  leaky P mode, 138
Branch cuts, 42, 43
  fluid-filled borehole models, 80
  integration, 49, 61–64
Bulk density, 7
Bulk modulus, 193

## C

Cased boreholes, 147–188
  attenuation measurements in, 179–184
    partition coefficients, 182, 183
    synthetic example, 183, 184
  boundary conditions for poorly bonded
    cases, 187, 188
  construction, 147–151
  D matrix elements, 185, 186